Lecture Notes in Computer Science 3004

Commenced Publication in 1973
Founding and Former Series Editors:
Gerhard Goos, Juris Hartmanis, and Jan van Leeuwen

Springer
*Berlin
Heidelberg
New York
Hong Kong
London
Milan
Paris
Tokyo*

Jens Gottlieb   Günther R. Raidl (Eds.)

Evolutionary Computation in Combinatorial Optimization

4th European Conference, EvoCOP 2004
Coimbra, Portugal, April 5-7, 2004
Proceedings

 Springer

Volume Editors

Jens Gottlieb
SAP AG
Neurottstr. 16, 69190 Walldorf, Germany
E-mail: jens.gottlieb@sap.com

Günther R. Raidl
Vienna University of Technology
Institute for Computer Graphics and Algorithms
Favoritenstraße 9-11/186, 1040 Vienna, Austria
E-mail: raidl@ads.tuwien.ac.at

Coverillustration: "Embrace" by Anargyros Sarafopoulos
http://ncca.bournemouth.ac.uk/main/staff/Anargyros/
Anargyros Sarafopoulos is a lecturer in computer animation and visualisation at the
National Centre for Computer Animation at Bournemouth University, where he applies genetic programming to the procedural representation of regular textures and
images using graph grammars and iterated function systems (IFS)

Library of Congress Control Number: 2004102324

CR Subject Classification (1998): F.1, F.2, G.1.6, G.2.1, G.1

ISSN 0302-9743
ISBN 3-540-21367-8 Springer-Verlag Berlin Heidelberg New York

Springer-Verlag is a part of Springer Science+Business Media

springeronline.com

© Springer-Verlag Berlin Heidelberg 2004
Printed in Germany

Typesetting: Camera-ready by author, data conversion by PTP-Berlin, Protago-TeX-Production GmbH
Printed on acid-free paper   SPIN: 10993279   06/3142   5 4 3 2 1 0

Preface

Evolutionary Computation (EC) involves the study of problem solving and optimization techniques inspired by principles of natural evolution and genetics. EC has been able to draw the attention of an increasing number of researchers and practitioners in several fields. Evolutionary algorithms have in particular been shown to be effective for difficult combinatorial optimization problems appearing in various industrial, economic, and scientific domains.

This volume contains the proceedings of EvoCOP 2004, the 4th European Conference on Evolutionary Computation in Combinatorial Optimization. It was held in Coimbra, Portugal, on April 5–7, 2004, jointly with EuroGP 2004, the 7th European Conference on Genetic Programming, and EvoWorkshops 2004, which consisted of the following six individual workshops: EvoBIO, the 2nd European Workshop on Evolutionary Bioinformatics; EvoCOMNET, the 1st European Workshop on Evolutionary Computation in Communications, Networks, and Connected Systems; EvoHOT, the 1st European Workshop on Hardware Optimisation; EvoIASP, the 6th European Workshop on Evolutionary Computation in Image Analysis and Signal Processing; EvoMUSART, the 2nd European Workshop on Evolutionary Music and Art; and EvoSTOC, the 1st European Workshop on Evolutionary Algorithms in Stochastic and Dynamic Environments.

EvoNet, the European Network of Excellence in Evolutionary Computing, organized its first events in 1998 as a collection of workshops that dealt with both theoretical and application-oriented aspects of EC. EuroGP, one of these workshops, soon became the main European conference dedicated to Genetic Programming. EvoCOP, held annually as a workshop since 2001, followed EuroGP. Due to its continuous growth in the past, EvoCOP became a conference in 2004, and it is now the premier European event focusing on evolutionary computation in combinatorial optimization. The events gave researchers and practitioners an excellent opportunity to present their latest research and to discuss current developments and applications, besides stimulating closer future interaction between members of this scientific community. Accepted papers of previous EvoCOP events were also published by Springer in the series Lecture Notes in Computer Science (LNCS Volumes 2037, 2279, and 2611).

EvoCOP 2004 covers evolutionary algorithms as well as related metaheuristics like memetic algorithms, ant colony optimization, and scatter search. The papers deal with representations, operators, search spaces, adaptation, comparison of algorithms, hybridization of different methods, and theory. The list of studied combinatorial optimization problems includes on the one hand classical academic problems such as graph coloring, network design, cutting, packing, scheduling, timetabling, the traveling salesperson problem, and vehicle routing, and on the other hand specific real-world applications.

The success of EvoCOP, so far the only international series of events specifically dedicated to the application of evolutionary computation and related methods to combinatorial optimization problems, is documented by a steadily growing number of submissions; see the table below. EvoCOP 2004 is the largest event of its series. A double-blind reviewing process made a strong selection among the submitted papers, resulting in an acceptance rate of 26.7%. All accepted papers were presented orally at the conference and are included in this proceedings volume. We would like to give credit to the members of the program committee, to whom we are very grateful for their quick and thorough work.

EvoCOP	submitted	accepted	acceptance ratio
2001	31	23	74.2%
2002	32	18	56.3%
2003	39	19	48.7%
2004	86	23	26.7%

EvoCOP 2004 was sponsored, for the last time, by EvoNet, whose activity as an EU-funded project has come to an end with the organization of this year's events. However, the figures reported above show that EvoCOP, as well as EuroGP and the EvoWorkshops, have reached a degree of maturity and scientific prestige that will allow the activity promoted by EvoNet during the last six years to go on, and presumably further expand in the years to come.

The organization of the event was made possible thanks to the active participation of several members of EvoNet. In particular, we want to thank Jennifer Willies, EvoNet's administrator, for her tremendous efforts.

April 2004   Jens Gottlieb
                                                                                                Günther R. Raidl

Organization

EvoCOP 2004 was organized jointly with EuroGP 2004 and the EvoWorkshops 2004.

Organizing Committee

Chairs: Jens Gottlieb, SAP AG, Germany

                 Günther R. Raidl, Vienna University of Technology, Austria

Local Chair: Ernesto Costa, University of Coimbra, Portugal

Program Committee

Hernan Aguirre, Shinshu University, Japan
Jarmo Alander, University of Vaasa, Finland
M. Emin Aydin, University of the West of England, UK
Jean Berger, Defence Research and Development Canada, Canada
Christian Bierwirth, University Halle-Wittenberg, Germany
Jürgen Branke, University of Karlsruhe, Germany
Edmund Burke, University of Nottingham, UK
David W. Corne, University of Exeter, UK
Ernesto Costa, University of Coimbra, Portugal
Carlos Cotta, University of Malaga, Spain
Peter Cowling, University of Bradford, UK
Bart Craenen, Vrije Universiteit, Amsterdam, The Netherlands
David Davis, NuTech Solutions, Inc., MA, USA
Marco Dorigo, Université Libre de Bruxelles, Belgium
Karl Dörner, University of Vienna, Austria
Anton V. Eremeev, Omsk Branch of the Sobolev Institute of Mathematics, Russia
David Fogel, Natural Selection, Inc., CA, USA
Bernd Freisleben, University of Marburg, Germany
Jens Gottlieb, SAP AG, Germany
Michael Guntsch, University of Karlsruhe, Germany
Jin-Kao Hao, University of Angers, France
Emma Hart, Napier University, UK
Jano van Hemert, CWI, The Netherlands
Jörg Homberger, Stuttgart University of Cooperative Education, Germany
Li Hui, University of Essex, UK
Mikkel T. Jensen, University of Aarhus, Denmark
Bryant A. Julstrom, St. Cloud State University, MN, USA

Sponsoring Institutions

- EvoNet, the Network of Excellence in Evolutionary Computing
- University of Coimbra, Coimbra, Portugal

Table of Contents

EvoCOP Contributions

Mutation Multiplicity in a Panmictic Two-Strategy Genetic Algorithm

Adnan Acan

Eastern Mediterranean University, Computer Engineering Department
Gazimagusa, T.R.N.C., Via Mersin 10, Turkey
adnan.acan@emu.edu.tr

Abstract. Fitness based selection procedures leave majority of population individuals idle, that is, they don't take place in any recombination operation although some of them have above average fitness values. Based on this observation, a two-phase two-strategy genetic algorithm using a conventional strategy with multiple mutation operators in the first phase is proposed. In the second phase, those individuals that are not sufficiently recombined in the first phase are reconsidered within a second strategy and recombined using multiple mutation operators only. In the second strategy, mutation operator probabilities are adaptively determined based on the cumulative fitness-gain achieved by each mutation operator over a number of generations. The proposed genetic algorithm paradigm is used for the solution of hard numerical and combinatorial optimization problems. The results demonstrate that the proposed approach performs much better than the conventional implementations in terms of solution quality and the convergence speed.

1 Introduction

Genetic algorithms (GA's) are biologically inspired search procedures that have been used successfully for the solution of hard numerical and combinatorial optimization problems. While successful applications of the standard implementation are presented in many scientific and industrial fields, there also has been a great deal on the derivation of various algorithmic alternatives towards faster and better localization of optimal solutions. In order to determine the most efficient ways of using GA's, many researchers have carried out extensive studies to understand several aspects such as the role and types of selection mechanisms, types of chromosome representations, types and application strategies of the genetic operators, parallel implementations, and hybrid algorithms [1], [2], [3].

This paper introduces a novel genetic algorithms strategy in which a population is evolved using a GA in which two different strategies are implemented in two consecutive phases. In the implementation of the proposed approach, it is implicitly assumed that the population is actually divided into two subpopulations; high fitness individuals that are frequently recombined during the first phase are stored within one subpopulation, whereas the individuals for which

J. Gottlieb and G.R. Raidl (Eds.): EvoCOP 2004, LNCS 3004, pp. 1–10, 2004.

the number of recombinations is below a predefined threshold are members of the second subpopulation. The main motivation and ideas leading the development of the proposed two-phase two-strategy GA are based on the following observation: due to selection procedures favoring individuals with higher survival capacity for recombination, most of the recombined individuals have fitness values above the population average. As it is also illustrated in the next section, only a very small percentage of individuals having below-average fitness enters the mating pool. Hence, the genetic algorithm steps are actually applied on an elite subset of the population, while individuals outside this subset are generated and wasted without being touched. Then, one may simply think of a population as if it is partitioned into two subpopulations implicitly created by fitness-based selection procedures. Since average fitness and ratio of above average individuals change over generations, the sizes of subpopulations are not fixed.The first phase of the proposed approach includes the application of a conventional GA with a single crossover operator and multiple mutation operators on the whole population. Then, the number of recombination operations performed on each parent individual is considered, and elements of the subpopulation to be processed in the second phase are determined. Since the majority of individuals within this subpopulation have below-average fitness and, hence, do not carry good solution properties, crossover operations between them will most probably not result in fitness improvements. However, one can make use of these individuals to provide further genetic diversity within a population by mutating them with multiple mutation operators.

The idea of using multiple mutation operators is an inspiration from nature; it is known that several mutation mechanisms such as point mutations, insertions, deletions, inversion, etc. occur simultaneously in nature [4]. In both phases, selection of mutation operators is based on a performance-directed scheme in which effectiveness of each operator is observed over a number of generations and operators with higher total fitness gain have higher probabilities of being selected. In the first phase, a selected mutation operator generates only one offspring with its associated small mutation probability, whereas, in the second phase, each selected mutation operator is allowed to generate more than one offspring based on its effectiveness. This is because, only mutation operators are used in the second phase. In this respect, currently the most effective mutation operator (the one with highest fitness gain) is applied on a parent for K_m times, where K_m is a predefined integer constant, whereas the least effective one is used to generate a single offspring only. The numbers of applications for the other mutation operators are determined with linear scaling. During the two phases, fitness values of the offspring individuals are computed and the mutation operator probabilities are updated based on these fitness values. A new generation begins with the first phase where the new panmictic population is the combination of individuals within the two offspring subpopulations generated in the current two phases. The implementation details of the proposed two-phase two-strategy GA approach are presented in Section 3.

This paper is organized as follows. The statistical reasoning behind the two-phase approach is introduced in Section 2. Section 3 covers the implementation details of the two-phase two-strategy GA paradigm. Experimental framework and obtained results are demonstrated in Section 4. Finally, conclusions and future research directions are specified.

2 Statistical Reasoning

The roulette-wheel and the tournament selection methods are the two most commonly used selection mechanisms in genetic algorithms. Both of these selection methods are fitness-based and they aim to produce more offspring from individuals having higher survival capacity. However, these selection operators leave a significant number of individuals having close to average fitness value unused in the sense that these individuals don't take part in any recombination operation.

In order to understand the above reasoning more clearly, let's take the minimization problem for the Ackley's function of 20 variables. A genetic algorithm with 200 individuals, uniform crossover with a crossover rate 0.7 and a point-mutation operator with probability 0.1 is considered. Since it is more commonly used, the tournament selection operator is selected for illustration. Statistical data are collected over 1000 generations. First, the ratio of the unused individuals to population size is shown in Figure 1. Obviously, on the average, 67% of the individuals remain unused in all generations, they are simply discarded and replaced by the newly produced offspring. This ratio of unused individuals is independent of the selection and encoding methods used; however, it changes with the population size. For example, this ratio is 56% for 100 individuals and 71% for 400 individuals.

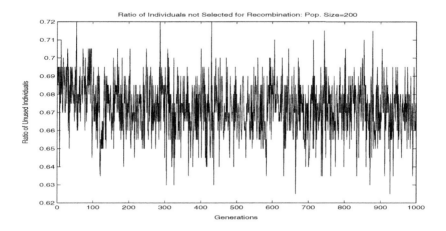

Fig. 1. The ratio of unselected individuals for a population size = 200

A more clear insight can be obtained from the ratio of unused individuals having a fitness value greater than the population's average fitness. As illustrated in Figure 2, on the average, 64% of the individuals having a fitness value above the population average are not used at all in any recombination operation.

Based on the above observations, the proposed approach is based on the following idea: a significant percentage of these idle individuals can be used to extend the genetic variance within the population. That is, when these individuals are mutated with multiple and adaptively selected mutation operators, it is possible to provide additional exploration over the solution space as exemplified and demonstrated by experimental evaluations in Section 4. In addition, while intensification is implemented within a subpopulation of superior individuals, pure diversification is achieved within another subpopulation through multiple adaptively selected mutation operators.

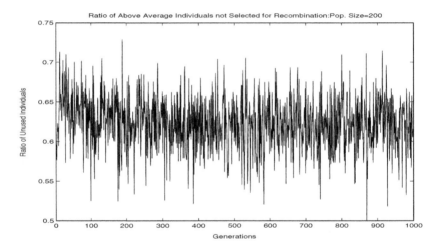

Fig. 2. The ratio of unselected individuals with above average fitness for a population size $= 200$

3   The Proposed Approach

Let's recall that, the main idea behind the proposed approach is to use two consecutive GA strategies to evolve a population which is implicitly divided into two subpopulations by fitness-based selection procedures. A conventional GA strategy empowered with multiple mutation operators is used to evolve a subpopulation of those individuals which are most frequently selected by a fitness-based selection procedure, whereas the subpopulation containing almost idle individuals is evolved with a mutation-only GA strategy. Mutation operators to be used are predefined and indexed using a mutation operators' index vector ($MOIV$). After initializing the population, $P(0)$, and finding the fitness values of individuals using the fitness function $F(.)$, mutation probabilities to be used in both phases are initialized to $1/|MOIV|$ for all mutation operators. There are two mutation operators' credit tables, $P1\_MOCT$ and $P2\_MOCT$, associated with the two phases of the algorithm. Both tables are $Kx|MOIV|$ matrices where rows represent the latest K generations while each column corresponds to a mutation operator. In a mutation operation, the gain achieved by the mutation operator is computed as the fitness improvement between the parent and its offspring. That is, if p is a parent and o is its offspring after mutation, then $max(F(o) - F(p), 0)$ is the improvement achieved by the corresponding mutation operator. Over one generation, these improvements are summed up to find the total improvement achieved by a particular mutation operator. $P1\_MOCT$ keeps these values for all mutation operators over the latest K generations in the first phase whereas $P2\_MOCT$ is the corresponding table used in the second phase. Hence, $P1\_MOCT(i, j)$ or $P2\_MOCT(i, j)$ is equal to the sum of fitness improvements achieved by the use of mutation operator j in the latest $i - th$ generation. As explained below, the contents of these tables are used to compute the mutation operator probabilities in the corresponding GA phases.

During the first phase of the algorithm, the number of times each individual is recombined is recorded in a table Rec_Table (recombination table). At the end of this first phase, those individuals for which the number of applied recombination operations is below a certain predetermined value are reconsidered and evolved with a different GA strategy employing multiple mutation operators only. In both of the strategies, mutation operator probabilities are not fixed and dynamically changed based on the cumulative fitness gain achieved by each mutation operator over a number of previous generations. In fact, since most of the crossover operators are applied on superior individuals, the proposed approach implements intensification over the subpopulation of superior individuals, whereas the inferior individuals are evolved by multiple mutation operators mainly for diversification purposes. In this respect, inferior individuals are allowed to produce more offspring compared to the superior individuals and only a best subset of them are taken and inserted in the next generation. Figure 3 illustrates the two-phase two-strategy GA framework, with mutation multiplicity, followed in the implementation of the proposed algorithm.

1. Generate initial population, $P(0)$.
2. Evaluate initial population, $F(P(0))$.
3. $P1\_MProb(i) = 1/|MOIV|, i = 1, \ldots, |MOIV|$; % Phase 1 mut. probabilities.
4. $P2\_MProb(i) = 1/|MOIV|, i = 1, \ldots, |MOIV|$; % Phase 2 mut. probabilities.
5. Initialize $P1\_MOCT$; % Init. phase 1 mut. operators credit table.
6. Initialize $P2\_MOCT$,; % Init. phase 2 mut. operators credit table.
7. ITER=1; DONE=FALSE.
8. while (NOT DONE)
9.   Set $Rec\_Count(i) = 0; i = 1, \ldots, |P(ITER - 1)|$.
10.   Compute $P1\_MProb(i), i = 1, \ldots, |MOIV|$.
11. $P1\_MOCT(1 + mod(ITER, K), j) = 0, j = 1, \ldots, |MOIV|$.
12. $Ind\_Count = 0$.

13.   while $Ind\_Count < Pop\_Size$
14. $\{p1, p2\} = Selection(P(ITER - 1))$.
15. $\{o1, [o2]\} = Crossover(p1, p2)$.
16. $Rec\_Count(Index(p1)) = Rec\_Count(Index(p1)) + 1$.
17. $Rec\_Count(Index(p2)) = Rec\_Count(Index(p2)) + 1$.
18. $Mutation = Select(P1\_MProb)$.
19. $Mutation(\{o1, [o2]\})$.
20. $Evaluate(\{o1, [o2]\})$.
21.   Insert $(\{o1, [o2]\})$ into $P(ITER)$.
22.   Update $P1\_MOCT(1 + mod(ITER, K), Mutation)$.
23. $Ind\_Count = Ind\_Count + |\{o1, [o2]\}|$.
24.   endwhile
25.   Inferiors $= \{P(i) | Rec\_Count(i) < T\}$.
26.   Compute $P2\_MProb(i), i = 1, \ldots, |MOIV|$.
27. $P2\_MOCT(1 + mod(ITER, K), j) = 0; j = 1, \ldots, |MOIV|$.
28.   for i=1 to $|Inferiors|$.
29. $Mutation = Select(MOIV, P2\_MProb)$.
30.   Compute $Offspring\_Count$ of $Mutation$.
31.   for j=1 to $Offspring\_Count$
32. $Offspring = Mutation(Inferiors(i))$.
33.   Evaluate $Offspring$.
34. $Fitness\_Gain = max(F(Offspring) - F(Inferiors(i)), 0)$.
35.   Insert($Mutated\_Inferiors, Offspring$).
36.   Update $P2\_MOCT(1 + Mod(ITER, K), Mutation, Fitness\_Gain)$.
37.   endfor
38.   endfor
39.   Sort $Mutated\_Inferiors$ in Ascending order of their fitness.
40.   Merge Superiors and Inferiors and keep only the best —P—.
41.   ITER=ITER+1.
42. endwhile

Fig. 3. Multiple crossover GA strategy.

The multiplicity feature within the proposed GA strategy is demonstrated within the two consecutive phases by mutation selection and operator fitness evaluation procedures. In steps 10 and 26, mutation operators' selection probabilities are computed proportional to their fitness values within the corresponding credit assignment tables as,

$$P1[2]\_MProb(i) = \sum_{j=1}^{K} P1[2]\_MOCT(j, i) / \sum_{l=1}^{K} \sum_{m=1}^{|MOIV|} P1[2]\_MOCT(l, m) \quad (1)$$

which divides the sum of improvements achieved by mutation operator i over the latest K generations by the sum of improvements achieved by all mutation operators over the same period. The selection procedure followed is exactly the method followed in roulette wheel selection. For the first phase, the mutation application probabilities are restricted to interval $[0.01, 0.1]$. That is, the mutation operator with minimum total fitness gain will be applied with probability 0.01 whereas the one with the highest total fitness gain is applied with probability 0.1. These values are experimentally determined and linear scaling is used to find the application probabilities of other mutation operators. In the second phase, the mutation application probabilities are restricted to $[0.25, 0.75]$ and applied in the same way of the first phase.

In the initialization of the operator credit tables, $P1[2]\_MOCT$, all mutation operators are assigned with the same selection probability, $P1[2]\_MProb(i) = 1/|MOIV|, i = 1, \cdots, |MOIV|$. Then, starting with $P(0)$, K generations are carried out to fill in the tables as explained above.

At the beginning of each generation, $P1[2]\_MOCT(1 + mod(ITER, K), j), j = 1, \cdots, |MOIV|$ is set to zero to record the fitness improvements achieved by each mutation operator during the current generation. Since the last K iterations are considered, $1 + mod(ITER, K)$ gives the row to be updated using offspring individuals within $P(ITER + 1)$.

Experimentally evaluated parameter values resulting in the best performance are given in the next section.

4   Experimental Work

4.1   Experimental Setup

Two test cases are considered for performance evaluation. The first one includes a number of benchmark problems from numerical optimization whereas the second one is a well-known instance of combinatorial optimization, the frequency assignment problem (FAP).

In all experiments, real-valued chromosomes are used for problem representations. The selection method used is the tournament selection with elitism. The elite size is 10% of the population size. The uniform crossover operator is employed with a crossover probability equal to 0.65 and the fixed mutation probability in conventional implementation is 0.05. Experiments are carried out using a population size of 200 individuals for the proposed approach. For the purpose of fair comparisons, the number of individuals generated at each iteration of conventional GA is made equal to 200+ number of fitness evaluations during the second stage; however only the best 200 of them are used to form the next population. This way, the total number of fitness evaluations is made the same for both algorithms. Each experiment is performed 10 times. All the tests were run over 1000 generations.

Five different mutation operators are used, namely, point-mutation, inversion, non-reciprocal translocation, gene-segment swapping, and gene-segment reinitialization. Point mutation and inversion are very well-known and widely used in practice. In non-reciprocal translocation, a randomly located and arbitrary-length chromosomal segment is relocated into a randomly selected location within the same chromosome. In gene-segment swapping, two randomly located equal-length chromosomal segments are swapped and in gene-segment reinitialization a randomly selected chromosomal segment is initialized with randomly chosen genes.

In the formation of the inferiors' subpopulation, those individuals for which the total recombination count after the first phase is less than or equal to 2% of the population size are selected as its elements. The maximum number of offspring that the best mutation operator generates is set to $K_m = 0.05 * Pop\_Size$.

4.2   Performance of the Proposed Approach in Numerical Optimization

Conventional implementation of GAs is compared with the proposed GA strategy for the minimization of a number of numerical optimization problems [5]. Each function has 20 variables. The best solution found using the conventional GAs and the proposed GA strategy are given in Table 1 where the proposed approach is named as GAMM (GAs with Mutation Multiplicity). GAMM provided very close to optimal results in all trials and outperformed the conventional GAs in the solution of numerical optimization problems in both solution quality and the convergence speed.

Table 1. Performance evalution of conventional GAs and GAMM for numerical optimization.

Function	Global Opt., n=Num. Vars.	Best Found: Conv. GA		Best Found: GAMM	
		Global Min.	ITER	Global Min.	ITER
Michalewicz	-9.66 ,$n = m = 10$	-8.55	500	-9.59	60
Griewangk	0, $n = 20$	0.0001	380	$1.0e^\blacksquare 18$	45
Rastrigin	0, $n = 20$	0.01	200	$1.0e^\blacksquare 15$	80
Schwefel	$-n * 418.9829, n = 20$	-8159	400	-8376	80
Ackley's	0, $n = 20$	0.0015	400	$1.0e^\blacksquare 18$	50
De Jong (Step)	0, n=20	1	500	0	30

4.3   Performance of the Proposed Approach in Combinatorial Optimization: The Frequency Assignment Problem

The frequency assignment problem (FAP) is defined as the assignment of frequencies within a predefined bandwidth to radio transmitters in a mobile telecommunication network in such a way that certain interference and traffic constraints are all satisfied. FAP is an NP-hard combinatorial optimization problem. In this paper, a well known minimum-order fixed-frequency assignment problem (MO-FFAP), namely the Philadelphia 21-cell MO-FAP, is considered and solved using the proposed GA strategy [6], [7]. This problem is handled by many authors in literature where two different demand vectors and several constraint matrices are used to define its different instances. The formulation and problem representations of MO-FFAP followed in our experiments can be found in [8].

Several commonly handled problem instances and the corresponding experimental results obtained for the Philadelphia system are listed in Table 2. For these test problem instances, the most difficult ones are the second and sixth instances. For all the other problems, the optimal solutions are found within a few seconds and within one or two generations. For the second and the sixth problem instances, best results are reported by [7] and the same results with the theoretical lower bounds are also reached by GAMM. Comparative convergence performances of GAMM and [7] are given in Table 3.

Table 2. Evaluation of GAMM with respect to published results

Problem	CM/CM(i,i)	Dem. Vec.	Low. Bound	Approach and the best solution						
				[7]	[6]	[8]	[11]	[10]	[9]	GAMM
1	$CM_1/5$	D_1	381	381	381	-	381	381	381	381
2	$CM_2/5$	D_1	427	427	-	-	433	463	-	427
3	$CM_1/7$	D_1	533	533	533	-	533	533	533	533
4	$CM_2/7$	D_1	533	533	533	-	533	533	533	533
5	$CM_2/5$	D_2	221	221	221	221	221	221	221	221
6	$CM_2/5$	D_2	253	253	-	268	263	275	-	253
7	$CM_1/7$	D_2	309	309	309	-	309	309	309	309
8	$CM_2/7$	D_2	309	309	309	309	309	309	309	309

* CM=Compatibility Matrix

Table 3. Evaluation of GAMM with respect to best published results

2nd Problem						6th Problem					
[7]			GAMM			[7]			GAMM		
$|F|$	Perf.	# Iters	$|F|$	Perf.	# Iters	$|F|$	Perf.	# Iters	$|F|$	Perf.	# Iters
			433	100%	50	259	100%	320	259	100%	20
432	100%	312	432	100%	50	258	75%	572	258	100%	20
431	80%	547	431	100%	70	257	50%	871	257	100%	80
430	70%	970	430	100%	110	256	30%	1064	256	100%	200
429	56%	2650	429	100%	200	255	10%	2110	255	80%	350
428	39%	3612	428	80%	360	254	4%	1164	254	60%	500
427	21%	3766	427	60%	700	253	2%	3420	253	50%	850

*$|F|$=Number of frequencies

5   Conclusions and Future Work

In this paper a novel GA paradigm employing a two-phase two-strategy GA with mutation multiplicity is introduced. The main ideas behind the proposed method are based on the observation that only a small elite subset of a population are used in recombination operations due to fitness-based selection operators. Hence, one can consider a population as if it is subdivided into two subpopulations which are evolved by two separate GA strategies. The overall algorithm works in two phases; in the first phase a conventional GA empowered with mutation multiplicity is used to evolve a single panmictic population. At the end of the first phase, those insufficiently utilized individuals are combined into a subpopulation that is evolved by a mutation only GA strategy.

The implemented algorithm is used to solve difficult numerical and combinatorial optimization problems and its performance is compared with that of conventional GAs. From the results of case studies, the proposed GAMM approach outperforms the conventional implementation in all trials. Additionally, the GAMM approach exhibits much better performance for both numerical and combinatorial optimization problems. In fact, problems from these classes are purposely chosen to examine this side of the proposed strategy.

This work can be further developed to test the effectiveness of mutation multiplicity in other algorithm development fields such as genetic programming, evolutionary programming, evolution strategies, and gene expression programming. In addition, mutation multiplicity in parallel implementations may also be a part of future studies.

References

1. Back T.: Evolutionary Algorithms in Theory and Practice, Oxford University Press, 1996.
2. Gen M. and Runwei C.: Genetic Algorithms in Engineering Design, John Wiley & Sons. Inc., 1997.
3. Miettinen K., Neitaanmaki P., Makela M.M., and Periaux J.: Evolutionary Algorithms in Engineering and Computer Science, John Wiley & Sons Ltd., 1999.
4. Russel P.J.: Genetics, 5th edition, Addison-Wesley 1998.
5. http://www.f.utb.cz/people/zelinka/soma/func.html.
6. Funabiki N., Takefuji Y.: A Neural network parallel algorithm for channel assignment problem in cellular radio networks, IEEE Trans. On Vehicular Technology, Vol. 41, No.4, p. 430-437, 1992.
7. Beckmann D., Killat U.: A new strategy for the application genetic algorithms to the channel assignment problem, IEEE Trans. On Vehicular Technology, Vol. 48, No. 4, p. 1261-1269, 1999.
8. Ngo C.Y., Li V.O.K.: Fixed channel assignment in cellular networks using a modified genetic algorithm, IEEE Trans. On Vehicular Technology, Vol. 47, No. 1, p. 163-172, 1998.
9. Chakraborty G.: An efficient heuristic algorithm for channel assignment problem in cellular radio networks", IEEE Trans. On Vehicular Technology, Vol. 50, No. 6, p. 1528-1539, 2001.
10. Battiti R.: A randomized saturation degree heuristic for channel assignment in cellular radio networks, IEEE Trans. On Vehicular Technology, Vol. 50, No.2, p. 364-374, 2001.
11. Wang W., Rushforth C.K.: An adaptive local search algorithm for the channel assignment problem", IEEE Trans. On Vehicular Technology, Vol. 45, No.3, p. 459-466, 1996.

Solving the Vehicle Routing Problem by Using Cellular Genetic Algorithms

Enrique Alba[1] and Bernabé Dorronsoro[2]

[1] Department of Computer Science, University of Málaga, `eat@lcc.uma.es`
[2] Central Computer Services, University of Málaga, `dorronsoro@uma.es`

Abstract. Cellular Genetic Algorithms (cGAs) are a subclass of Genetic Algorithms (GAs) in which the population diversity and exploration are enhanced thanks to the existence of small overlapped neighborhoods. Such a kind of structured algorithms is specially well suited for complex problems. In this paper we propose the utilization of some cGAs with and without including local search techniques for solving the vehicle routing problem (VRP). A study on the behavior of these algorithms has been performed in terms of the quality of the solutions found, execution time, and number of function evaluations (effort). We have selected the benchmark of Christofides, Mingozzi and Toth for testing the proposed cGAs, and compare them with some other heuristics in the literature. Our conclusions are that cGAs with an added local search operator are able of always locating the optimum of the problem at low times and reasonable effort for the tested instances.

1 Introduction

Transportation is a major problem domain in logistics, and so it represents a substantial task in the activities of many companies. In some market sectors, transportation means a high percentage of the value added to goods. Therefore, the utilization of computerized methods for transportation often results in significant savings ranging from 5% to 20% in the total costs, as reported in [1]. A distinguished problem in the field of transportation consists in defining the optimal routes for a fleet of vehicles to serve a set of clients. In this problem an arbitrary set of clients must receive goods from a central depot. This general scenario presents many chances for several concrete definitions of subclasses of problems: determining the optimal number of vehicles, finding the shortest routes, etc. all this subject to many restrictions like vehicle capacity, time windows for deliveries, etc. This variety of scenarios leads to a plethora of problem variants in practice.

The basic problem class we address in this paper is known as the Vehicle Routing Problem (VRP) [2], which consists in delivering goods to a set of customers with known demands through minimum-cost vehicle routes originating and terminating at the depot, as seen in Fig. 1 –for a more detailed description of the problem see Sect. 2.

The VRP is an important class of problem, since solving it is equivalent to solving multiple TSP problems at once, and thus it has been studied both theo-

J. Gottlieb and G.R. Raidl (Eds.): EvoCOP 2004, LNCS 3004, pp. 11–20, 2004.

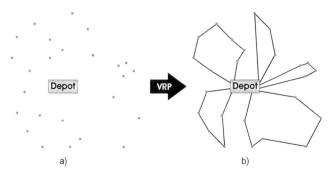

Fig. 1. The Vehicle Routing Problem finds the minimum cost routes –b)– to serve a set of customers (points) geographically distributed, from a depot –a)–

retically and in practice [3]. There is a large number of extensions to the canonical VRP, but in this paper, we address the base problem without time windows.

There has been a steady evolution in the design of solution methodologies in the past, both exact and approximate, for this problem. However, no known exact method is capable of consistently solving to optimality instances involving more than 50 customers [1]. Instead, heuristics are mostly preferred.

The contribution of this work is to define a powerful yet simple Cellular Genetic Algorithm (cGA) [4] that is able to compete with the best known approaches in the resolution of the VRP in terms of the solution found, the execution time, and the number of evaluations made. In addition, it represents a kind of paradigm much simpler of customization and comprehension than others such as Tabu Search (TS) or very specialized hybrid algorithms. Local search techniques can be quite naturally embedded into a cGA, and we test here to which extent it can be advantageous. We test our algorithms over a selection of the Christofides et al. [3] benchmark instances, and compare our results with some other known heuristics.

The paper is arranged in the following manner. Section 2 gives a mathematical formulation for the VRP. Our cGA is described in Sect. 3. Section 4 presents the results of our algorithms, and compares them with those of some heuristics in the literature. The conclusions and some future work lines are presented in Sect. 5.

2   Vehicle Routing Problem (VRP)

The VRP is a well known integer programming problem which falls into the category of NP-hard problems [5]. It is defined on an undirected graph $G = (V, E)$ where $V = \{v_0, v_1, \ldots, v_n\}$ is a vertex set and $E = \{(v_i, v_j)/v_i, v_j \in V, i < j\}$ is an edge set. The depot is represented by vertex v_0, and it is from where that m identical vehicles of capacity Q must service all the cities or customers, represented by the set of n vertices $\{v_1, \ldots, v_n\}$. We define on E a non-negative cost, distance or travel time matrix $C = (c_{ij})$ between customers v_i and v_j. Each customer v_i has non-negative demand q_i and drop time δ_i (time needed to unload all goods). Let be $R_1, \ldots, R_m$ a partition of V representing the routes of the vehicles to service all the customers. The cost of a given route $(R_i = \{v_0, v_1, \ldots, v_{k+1}\})$, where $v_j \in V$ and $v_0 = v_{k+1} = 0$ (0 denotes the depot), is given by:

$$\text{Cost}(R_i) = \sum_{j=0}^{k} c_{j,j+1} + \sum_{j=0}^{k} \delta_j \ , \tag{1}$$

and the cost of the problem solution (S) is:

$$F_{\text{VRP}}(S) = \sum_{i=1}^{m} \text{Cost}(R_i) \ . \tag{2}$$

The VRP consists in determining a set of a maximum of m routes (i) of minimum total cost (see (2)); (ii) starting and ending at the depot v_0; and such that (iii) each customer is visited exactly once by exactly one vehicle; (iv) the total demand of any route does not exceed Q ($\sum_{v_j \in R_i} q_j \leq Q$); and (v) the total duration of any route is not larger than a preset bound D ($\text{Cost}(R_i) \leq D$). The number of vehicles is either an input value or a decision variable. In this study, the length of routes is minimized independently of the used number of vehicles.

3   Algorithms for Solving the VRP

In this section we discuss the algorithms we plan to use later in the paper to solve the VRP. Initially we present a cGA with structured population as the basic tool for the search. In addition, since there exists a long tradition in VRP to include local search techniques to improve partial solutions; and, since in cGAs it is quite direct to include such kind of procedures, we will further discuss in this section several mechanisms to perform local optimization: 2-Opt and λ-Interchange.

3.1   Cellular Genetic Algorithms for the VRP

Genetic Algorithms (GAs) are based on an analogy with biological evolution. A population of individuals representing tentative solutions is maintained. New individuals are produced by combining members of the population, and these replace existing individuals with some policy. Cellular GAs are a subclass of GAs in which the population is structured in a specified topology, so that individuals may only interact with their neighbors (see Fig. 2). These overlapped small neighborhoods help in exploring the search space because the induced slow diffusion of solutions through the population provides a kind of exploration, while exploitation takes place inside each neighborhood by genetic operations. In Algorithm 1 we can see a pseudocode of our basic cGA [6].

In this cGA, the population is structured in a 2D toroidal grid, and the neighborhood defined on it –line 5– always contains 5 individuals: the considered one (position(x,y)) plus the North, East, West, and South individuals (called

Algorithm 1 Pseudocode of cGA

```
 1: proc Steps_Up(cga)      //Algorithm parameters in 'cga'
 2: for s ← 1 to MAX_STEPS do
 3:    for x ← 1 to WIDTH do
 4:       for y ← 1 to HEIGHT do
 5:          n_list←Get_Neighborhood(cga,position(x,y));
 6:          parents←Local_Select(n_list);
 7:          aux_indiv←Recombination(cga.Pc,parents);
 8:          aux_indiv←Mutation(cga.Pm,aux_indiv);
 9:          aux_indiv←Local_Search(cga.Pl,aux_indiv);
10:          Evaluate_Fitness(aux_indiv);
11:          Insert_If_Better(position(x,y),aux_indiv,cga,aux_pop);
12:       end for
13:    end for
14:    cga.pop←aux_pop;
15:    Update_Statistics(cga);
16: end for
17: end_proc Steps_Up;
```

Panmictic GA

Cellular GA

Fig. 2. Panmictic and cellular populations

NEWS, linear5, or Von Neumann). Binary tournament selection is used in the neighborhood for the first parent, while the other one is the considered individual itself (Local_Select) (line 6). Genetic operators are applied to the individuals in lines 7 and 8 (explained in Sect. 3.1). We can add different local search techniques to our basic cGA by simply including line 9 to the algorithm (see Sect. 3.2). After applying these operators, we evaluate the fitness value of the new individual (line 10), and insert it on its equivalent place in the new population –line 11– only if its fitness value is larger than the parent's one (elitist replacement).

After applying the above mentioned operators to the individuals we replace the old population by the new one (line 14), and we calculate some statistics (line 15). It can be seen how in our algorithm new individuals replace old ones en bloc (synchronously) and not incrementally. The fitness value assigned to every individual is computed like follows:

$$f(\boldsymbol{S}) = F_{\text{VRP}}(\boldsymbol{S}) + \lambda \cdot \text{overcap}(\boldsymbol{S}) + \mu \cdot \text{overtm}(\boldsymbol{S}) \ , \tag{3}$$

$$f_{\text{eval}}(\boldsymbol{S}) = f_{\max} - f(\boldsymbol{S}) \ . \tag{4}$$

The objective of our algorithm is to maximize $f_{\text{eval}}(\boldsymbol{S})$ (4) by minimizing $f(\boldsymbol{S})$ (3). The value $f_{\max}$ must be larger or equal to that of the worst feasible solution for the problem. Function $f(\boldsymbol{S})$ is calculated by adding the total costs of all the routes ($F_{\text{VRP}}(\boldsymbol{S})$), and penalizes the fitness value if the capacity of any vehicle and/or the time of any route are exceeded. Functions overcap($\boldsymbol{S}$) and overtm($\boldsymbol{S}$) return the overhead in capacity and time, respectively, of the solution with respect to the maximum allowed value of each route. These values returned by overcap($\boldsymbol{S}$) and overtm($\boldsymbol{S}$) are weighted by multiplying them with the values λ and μ, respectively.

Representation of Individuals. In a GA, individuals represent candidate solutions. A candidate solution to an instance of the VRP must specify the number of vehicles required, the allocation of the demands to all these vehicles, and also the delivery order of each route. We adopted a representation consisting of a permutation of integers. Each permutation will contain both customers and route splitters (delimiting different routes), so we will use a permutation of numbers $[0 \ldots n-1]$ with length $n = c + k$ for representing a solution for the VRP with c customers and a maximum of $k + 1$ possible routes. Customers are represented with numbers $[0 \ldots c-1]$, while route splitters belong to the range $[c \ldots n-1]$.

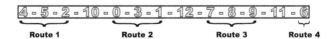

Route 1   Route 2   Route 3   Route 4

Fig. 3. Individual representing a solution

Each route is composed by the customers in the individual between two route splitters. For example, in Fig. 3 we plot an individual representing a possible solution for a VRP instance with 10 customers using 4 vehicles at most. Values $[0, \ldots, 9]$ represent the customers while $[10, \ldots, 12]$ are the route splitters. **Route 1** begins at the depot, visits customers 4–5–2 (in that order), and returns to the depot. **Route 2** goes from the depot to customers 0–3–1 and returns. The vehicle of **Route 3** starts at the depot and visits customers 7–8–9. Finally, in **Route 4**, only customer 6 is visited. This representation allows empty routes by simply placing two route splitters together without clients between them.

Genetic Operators. Two genetic operators are applied to all the individuals with preset probabilities along the evolution. These two operators are recombination and mutation. The recombination operator we use is an edge recombination operator (ERX) [7], that builds an offspring by preserving edges from both parents. The mutation operator used in the algorithm consists in applying Insertion, Swap or Inversion operations to each gene with equal probability (see Fig. 4). The Insertion [8] operator selects a customer and inserts it in another randomly selected place. Swap [9] consists in randomly selecting two customers and exchanging them. Finally, Inversion [10] reverses the visiting order of the customers between two randomly selected cut points. In all the three operators changes might occur in an intra or inter-route way.

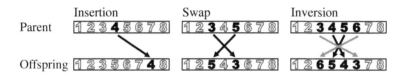

Fig. 4. The three different mutations used

3.2   Local Search

It is very clear after all the existing literature on VRP that local search is almost mandatory to achieve results of high quality [11] [12]. This is why we envision from the beginning the application of two of the most successful techniques in the past years. Then we will not only consider canonical cGAs for VRP, but also will add a local post-optimization step by applying 2-Opt [13] and/or λ-Interchange [14] local optimization methods to every individual after each generation.

The 2-Opt simple local search method works inside each route. It randomly selects two non-adjacent edges (i.e. (a, b) and (c, d)) of a single route, delete them, thus breaking the tour into two parts, and then reconnects those parts in the other possible way (i.e. (a, c) and (b, d)).

The λ-Interchange local optimization method we use is based on the interchange of all the possible combinations for up to λ customers between sets of routes. Hence, this method results in customers either being shifted from one route to another, or being exchanged between routes. The mechanism can be described as follows. A solution to the problem is represented by a set of routes $S = \{R_1 \ldots R_p \ldots R_q \ldots R_k\}$, where R_i is the set of customers serviced in route i. New neighboring solutions can be obtained by applying λ-Interchange between a pair of routes R_p and R_q by replacing each subset of customers $S_1 \subseteq R_p$ of size $|S_1| \leq \lambda$ with any other one $S_2 \subseteq R_q$ of size $|S_2| \leq \lambda$. This way, we get two new routes $R'_p = (R_p - S_1) \bigcup S_2$ and $R'_q = (R_q - S_2) \bigcup S_1$, which are part of the new solution $S' = \{R_1 \ldots R'_p \ldots R'_q \ldots R_k\}$.

4   Experimentation

In this section we describe the experiments applying these ideas to a set of problems taken from the well-known OR library [15]. Our algorithms have been implemented in Java on a 2.4 GHz PC (Linux OS). They have been tested with instances C1, C2, C3, C6, C7, C8, C12, and C14 [3] (not enough room for the whole benchmark).

The first three instances contain data for 50, 75, and 100 customers respectively, with capacity restrictions on vehicles but not on the distance. Datasets C6, C7 and C8 are like C1, C2 and C3, respectively, with the additional restrictions of maximal route distance and the existence of a drop time. An additional feature of C12 with respect to the three first ones is that the 100 customers to visit are geographically clustered. The last instance included, C14 is really the C12 instance with the restrictions of maximal route distance and the addition of drop times.

We have used four different algorithms, all of them extracted from the metaheuristic sheet of a cGA described in Sect. 3. The difference among them is just the local search method used. The first algorithm studied does not use any local search method (cGA). The second one uses the 2-Opt method (cGA2o). The two last studied algorithms use 2-Opt plus a λ-Interchange (with a maximum of 20 search steps), and return the best individual found with any of the two methods on each local search step. These algorithms differ in the value of λ used. One of them implements $\lambda = 1$ (cGA2o1i), and the other one utilizes $\lambda = 2$ (cGA2o2i).

For each individual, the recombination probability is set to 0.65 in all of our algorithms. The mutation probability for an individual is 0.85, and its alleles are mutated with probability $1/L$, being L the length of the chromosome. The local search (if used) is applied to all individuals (probability 1.0). Population size is 100 individuals in any case, and we perform 30 independent runs to get reliable statistical results. For the evaluation of the fitness value, we have used values 10^7, 10^3 and 10^3 for parameters f_{max}, λ and μ, respectively.

Next, in Sect. 4.1, we present the results obtained with our proposed algorithms, while in Sect. 4.2 we compare our results with the existing literature.

4.1   Performance of the Proposed Algorithms

Let us now proceed to analyze the results. The proposed algorithms can easily be compared by consulting tables 1, 2 and 3, containing the best solution found (total route distance), and the average execution time and number of evaluations needed, respectively. The optimum values are bolded for the reader's commodity.

Our conclusions are quite stable and relatively clear. The results worked out by the canonical cGA (Table 1) are far from the best known results for the tested instances. However, just by including a local improvement with 2-Opt (cGA2o algorithm) the total cost of the solutions found in all the instances is greatly reduced with respect to cGA, although no optimal solution is still found. It is by adding the λ-Interchange local optimization that we achieve really interesting results, since we can solve all the instances to the optimum. Table 1 shows that cGA2o1i finds the best known solution for all the instances, having a similar behavior as for cGA2o2i (being this last only slightly less accurate for C3).

If we analyze average execution times (see Table 2, figures in seconds), then we can appreciate a clear increment as the complexity of the algorithms is enlarging: this sounds common sense. Therefore, the simpler the algorithm the smaller the execution time. In most cases times are quite affordable and reasonable.

Finally, Table 3 shows the average number of function evaluations made by the algorithm during the search. The effort for cGA and cGA2o is difficult to evaluate since it is an effort to lead the algorithm to a local optimum (most runs reach the maximum number of allowed evaluations without finding a solution). On the other hand, the numerical effort of cGA2o1i and cGA2o2i are quite similar and meaningful, indicating that cGA2o1i is preferable over cGA2o2i since it performs a search of similar accuracy in a smaller time.

As a final conclusion, we can point out cGA2o1i as being the more accurate algorithm showing very reasonable times and effort to locate the optimum. The other algorithms are either too explorative (inaccurate) or unnecessarily complex.

4.2   Comparison against Other Techniques

In this section we compare our algorithms against the best so far techniques reported in the literature, as any rigorous work should do. We have selected

Table 1. Best solution found

	cGA	cGA2o	cGA2o1i	cGA2o2i
C1	548.83	533.09	**524.61**	**524.61**
C2	907.53	874.41	**835.26**	**835.26**
C3	937.36	838.70	**826.14**	827.39
C6	568.83	558.99	**555.43**	**555.43**
C7	1030.23	973.48	**909.68**	**909.68**
C8	1053.82	903.66	**865.94**	**865.94**
C12	986.62	891.32	**819.56**	**819.56**
C14	1007.92	904.43	**866.37**	**866.37**

Table 2. Average time (in seconds)

	cGA	cGA2o	cGA2o1i	cGA2o2i	RT	Prins	BB
C1	3.53 ± 0.11	11.28 ± 0.63	42.38 ± 1.82	48.63 ± 1.98	11	< 1	120
C2	4.81 ± 0.11	23.45 ± 0.71	252.54 ± 16.26	289.06 ± 11.32	68	184	860
C3	7.86 ± 0.25	60.47 ± 2.27	516.74 ± 30.58	569.14 ± 29.76	900	497	1674
C6	3.54 ± 0.08	9.85 ± 0.31	47.68 ± 2.24	55.92 ± 6.92	—	106	140
C7	4.80 ± 0.08	20.02 ± 0.70	268.11 ± 12.95	312.08 ± 137.59	—	315	630
C8	7.93 ± 0.31	49.75 ± 2.10	628.16 ± 191.95	659.81 ± 28.02	—	664	303
C12	7.40 ± 0.49	53.21 ± 2.34	485.58 ± 82.60	536.42 ± 26.96	350	7	433
C14	7.87 ± 0.24	47.63 ± 2.01	583.54 ± 65.22	651.54 ± 32.26	—	591	284

Table 3. Average number of evaluations

	cGA	cGA2o	cGA2o1i	cGA2o2i
C1	100100.0 ± 0.0	213310.10 ± 1770.36	71725.73 ± 2257.59	72370.60 ± 2741.85
C2	120100.0 ± 0.0	373167.20 ± 4875.20	171668.90 ± 7818.45	170747.50 ± 5825.33
C3	150100.0 ± 0.0	481971.10 ± 6458.94	210094.10 ± 9350.17	208350.70 ± 8665.93
C6	100100.0 ± 0.0	211402.10 ± 1649.12	75029.00 ± 2729.36	75023.80 ± 2442.21
C7	120100.0 ± 0.0	365590.50 ± 3716.58	170737.40 ± 6055.16	171175.10 ± 5889.34
C8	150100.0 ± 0.0	475392.50 ± 6469.34	227187.80 ± 9307.24	223854.50 ± 9126.49
C12	150100.0 ± 0.0	488714.20 ± 8261.53	193967.70 ± 8113.52	193482.90 ± 6525.01
C14	150100.0 ± 0.0	486746.80 ± 7819.13	222168.50 ± 7604.63	221596.50 ± 9429.92

some classic algorithms, like savings [16], sweep [17], 1-petal or 2-petal [18], and some newer ones like tabu search TS, other GAs, and also ant systems. These three last techniques are known to perform very well on the VRP.

The TS algorithm included is that of Rochat and Taillard [11] (RT), the genetic algorithms are those of Prins [19], and Berger and Barkaoui [12] (BB), and, the two ant algorithms included are very recent works proposed by Bullnheimer et al. [20] (AS) and SavingsAnts, by Doerner et al. [21] (SA).

As shown in Table 4, RT, Prins, and cGA2o1i outperform the other algorithms, finding the best known solution for all the studied instances. Hence, the two mentioned ant algorithms perform with a lower accuracy than existing TS and GAs.

We must however point out a clear distinction: we are just hybridizing two canonical algorithms whose definitions are simple and extremely easy for customization on these and other similar problems. The same cannot be said for other highly specialized algorithms, that require a great customization effort from the researchers. Simplicity and clarity (and even potential extensibility) are a clear contribution of our algorithms with respect to the others.

In Table 2 we compare all the algorithms for which execution times have been reported. Times of RT are calculated by the authors on a Silicon Graphics Indigo –100 MHz– (no run-times were provided for C6, C7, C8 and C14), the genetic algorithm of Prins is implemented in Delphi on a 500 MHz PC under Windows 95, and BB is implemented in C++ on a 400 MHz Pentium processor. The algorithm of Prins is the globally faster for instances C1, C3 and C12, while RT is the fastest one just in the case of C2. The best times for instances C8 and C14 are hold by BB, and finally, our cGA2o1i is faster than the others for instances C6 and C7.

Table 4. Our best cGA vs. other heuristics and best known solutions

	Savings	Sweep	1-Petal	2-Petal	RT	AS	SA	Prins	BB	cGA2o1i
C1	578.56	531.90	531.90	**524.61**	**524.61**	**524.61**	**524.61**	**524.61**	**524.61**	**524.61**
C2	888.04	884.20	885.02	854.09	**835.26**	844.31	838.60	**835.26**	**835.26**	**835.26**
C3	878.70	846.34	836.34	830.40	**826.14**	832.32	838.38	**826.14**	827.39	**826.14**
C6	616.66	560.08	560.08	560.08	**555.43**	560.24	**555.43**	**555.43**	**555.43**	**555.43**
C7	974.79	965.51	968.89	922.75	**909.68**	916.21	913.01	**909.68**	**909.68**	**909.68**
C8	968.73	883.56	877.80	877.29	**865.94**	866.74	870.10	**865.94**	868.32	**865.94**
C12	824.42	919.51	824.77	824.77	**819.56**	**819.56**	**819.56**	**819.56**	**819.56**	**819.56**
C14	868.50	911.81	894.77	885.87	**866.37**	867.07	**866.37**	**866.37**	**866.37**	**866.37**

5   Conclusions and Further Work

This paper presents different algorithms based in cGAs for solving the vehicle routing problem. The used cellular GA technique maintains population diversity for more time with respect to panmictic (single population) GAs due to the use of small overlapped neighborhoods. This feature frequently prevents the algorithm from getting stuck in local optima.

First, an initial and simple approach to a cGA for solving the VRP has been proposed. It has been shown that adding a local search step to that simple approach is enough to get a really powerful algorithm. The local search step is made up of simple methods like 2-Opt or λ-Interchange. With them we have got the best known results for all the proposed instances, specially with cGA2o1i.

The algorithms proposed are compared against a TS algorithm able to find the best known results for all the instances of the CMT benchmark (proposed by Rochat and Taillard), to a genetic algorithm proposed by Prins, to a hybrid genetic algorithm (Berger et al.), and also to two ant algorithms, as well as against other ad hoc heuristics. It has been proved that cGA2o1i is faster than the algorithms compared to it in this paper for many instances.

In addition, the cGAs studied in this paper have the advantage of their simplicity and accuracy with respect to the existing techniques. Our algorithms are just a first approach, susceptible of improvement. Hence, as a future work, one can think about testing some other local search techniques, like using just λ-Interchange alone without 2-Opt. Another possible future work may be trying to better tune the cGA parameters, and test the algorithms on larger benchmarks.

Acknowledgements. This work has been funded by MCYT and FEDER under contract TIC2002-04498-C05-02 (the TRACER project) http://tracer.lcc.uma.es.

References

1. Toth, P., Vigo, D.: The Vehicle Routing Problem. Monographs on Discrete Mathematics and Applications. SIAM, Philadelphia (2001)
2. Dantzing, G., Ramster, R.: The truck dispatching problem. Management Science **6** (1959) 80–91
3. Christofides, N., Mingozzi, A., Toth, P.: The Vehicle Routing Problem. In: Combinatorial Optimization. John Wiley (1979) 315–338
4. Manderick, B., Spiessens, P.: Fine-grained parallel genetic algorithm. In Schaffer, J., ed.: 3rd ICGA, Morgan-Kaufmann (1989) 428–433
5. Lenstra, J., Kan, A.R.: Complexity of vehicle routing and scheduling problems. Networks **11** (1981) 221–227
6. Whitley, D.: Cellular genetic algorithms. In Forrest, S., ed.: Proceedings of the 5th ICGA, Morgan-Kaufmann, CA (1993) 658
7. Whitley, D., Starkweather, T., Fuquay, D.: Scheduling problems and traveling salesman: The genetic edge recombination operator. In Schaffer, J., ed.: 3rd ICGA, Morgan-Kaufmann (1989) 133–140
8. Fogel, D.: An evolutionary approach to the traveling salesman problem. Biological Cybernetics **60** (1988) 139–144
9. Banzhaf, W.: The "molecular" traveling salesman. Biological Cybernetics **64** (1990) 7–14
10. Holland, J.: Adaptation in Natural and Artificial Systems. University of Michigan Press, Ann Arbor, MI (1975)
11. Rochat, Y., Taillard, E.: Probabilistic diversification and intensification in local search for vehicle routing. J. of Heuristics **1** (1995) 147–167
12. Berger, J., Barkaoui, M.: A hybrid genetic algorithm for the capacitated vehicle routing problem. In Cantú-Paz, E., ed.: GECCO03. LNCS 2723, Illinois, Chicago, USA, Springer-Verlag (2003) 646–656
13. Croes, G.: A method for solving traveling salesman problems. Operations Research **6** (1958) 791–812
14. Osman, I.: Metastrategy simulated annealing and tabu search algorithms for the vehicle routing problems. Annals of Operations Research **41** (1993) 421–451
15. Beasley, J.: OR-library: Distributing test problems by electronic mail. J. of the Operational Research Society **11** (1990) 1069–1072
16. Clarke, G., Wright, J.: Scheduling of vehicles from a central depot to a number of delivery points. Operations Research **12** (1964) 568–581
17. Wren, A., Holliday, A.: Computer scheduling of vehicles from one or more depots to a number of delivery points. Operational Research Quarterly **23** (1972) 333–344
18. Ryan, D., Hjorring, C., Glover, F.: Extensions of the petal method for vehicle routing. J. of the Operational Research Society **44** (1993) 289–296
19. Prins, C.: A simple and effective evolutionary algorithm for the vehicle routing problem. Computers and Operations Research (2003) In Press, Corrected Proof.
20. Bullnheimer, B., Hartl, R., Strauss, C.: An improved ant system algorithm for the vehicle routing problem. Annals of Operations Research **89** (1999) 319–328
21. Reimann, M., Doerner, K., Hartl, R.: D-ants: Savings based ants divide and conquer the vehicle routing problem. Computers & Operations Res. **31** (2004) 563–591

Landscape Regularity and Random Walks for the Job-Shop Scheduling Problem

Christian Bierwirth[1], Dirk Christian Mattfeld[2], and Jean-Paul Watson[3]

[1] Martin-Luther-University of Halle-Wittenberg, Germany
`bierwirth@wiwi.uni-halle.de`
[2] University of Hamburg, Germany
`dirk@uni-bremen.de`
[3] Sandia National Labs, Albuquerque, USA
`jwatson@sandia.gov`

Abstract. We perform a novel analysis of the fitness landscape of the job-shop scheduling problem (JSP). In contrast to other well-known combinatorial optimization problems, we show that the landscape of the JSP is non-regular, in that the connectivity of solutions is variable. As a consequence, we argue that random walks performed on such a landscape will be biased. We conjecture that such a bias should affect both random walks and local search algorithms, and may provide a partial explanation for the remarkable success of the latter in solving the JSP.

1 Introduction

The job shop scheduling problem (JSP) is a notoriously difficult NP-hard combinatorial optimization problem [4,8] and is consequently used as one benchmark by which the effectiveness of local search algorithms is judged [14,18,2,3,13]. Several attempts have been undertaken to explain the difficulty of the JSP through investigation of problem structure [16,24,5]. These approaches are based on the notion of a fitness landscape, the structure of which is known to influence problem difficulty.

One approach to characterizing fitness landscape structure is to assess the ruggedness of landscape by measuring the auto-correlation of solution quality observed in a time-series generated by a random walk [11,25]. A prerequisite for this approach is the regularity of the search space, such that all elements of the landscape are visited by a random walk with equal probability. Many prominent optimization problems, including the traveling salesman problem [12,22], the quadratic assignment problem [23,17], and the flow shop scheduling problem [19] possess regular search spaces.

In this paper, we show that the configuration space underlying a fitness landscape can be non-regular, and that the JSP is a prominent example of such a combinatorial optimization problem. Further, we show experimentally that random walks performed on such a landscape will be biased. We conjecture that such a bias should affect both random walks and directed stochastic local search algorithms for the JSP.

J. Gottlieb and G.R. Raidl (Eds.): EvoCOP 2004, LNCS 3004, pp. 21–30, 2004.

This paper is organized as follows. In Section 2 we introduce a general model of combinatorial search spaces in order to describe regularity and drift more formally. In Section 3, we verify the irregularity of the JSP search space on a small problem instance, and consider the potential impact of the irregularity on the behavior of both a random walk and directed stochastic local search. In Section 4, we experimentally confirm the irregularity of the JSP search space on the well-known ft10 JSP instance. We conclude in Section 5 and discuss directions for further research.

2   A Model of Combinatorial Search Spaces

2.1   Configuration Spaces and Fitness Landscapes

Local search algorithms for combinatorial optimization problems aim to identify a solution with an optimal objective function value out of a finite but vast set of feasible solutions to a problem instance. The set of feasible solutions is structured by a neighborhood that defines the set of possible moves from a solution at each point during search. The configuration space of a problem is defined as graph $G = (V, E)$. Feasible solutions are represented by nodes $v \in V$ and a move between two feasible solutions $v_1, v_2 \in V$ is represented by an edge $(v_1, v_2) \in E$.

We make some assumptions regarding G. First, G is a connected graph such that a path exists between any pair of nodes v_1 and v_2. This property enables local search to potentially locate an optimal solution from an arbitrary starting point in the configuration space. Second, for simplicity, we assume G is undirected, i.e., edges are bidirectional such that every move can be reversed. The degree $d(v)$ of a node $v \in V$ is defined by the number of incident edges and represents the size of the move set at v. Every node $v \in V$ is assigned a value $f(v)$ from a real-valued objective function f leading to the node-weighted graph $\hat{G} = (V, E, f)$. Such a graph is also known as a fitness landscape [21].

Numerous researchers have analyzed the structure of the fitness landscape, primarily in an effort to shed light on the issue of problem difficulty for local search. A central measure of landscape structure is the fitness-distance correlation (FDC) [9,10]. Given a set of feasible solutions, FDC quantifies the correlation between the objective function value and the length of a path to an optimal solution in $\hat{G}$. Given an objective of minimization, the presence of significant FDC indicates that $\hat{G}$ exhibits a big valley structure [19,5]. Another important measure quantifies the ruggedness of a landscape, i.e., the difference in objective function values observed during a random walk on $\hat{G}$ is used to determine the autocorrelation length of the landscape.

In general, landscapes with a low fitness-distance correlation and a short autocorrelation length are expected to be difficult for local search, independent of the particular algorithm used. However, despite extensive analyses of the fitness landscapes of numerous combinatorial optimization problems, the link between problem difficulty and fitness landscape structure is poorly understood.

2.2   Connectivity, Regularity, and Drift

The deficiency of the proposed measures may be due to the neglect of problem-specific properties that cannot be observed at the surface of a fitness landscape $\hat{G}$ [20]. An important assumption made for meaningful statistics obtained from random walks is that a landscape looks (globally) the same everywhere [7]. From the viewpoint of the underlying configuration space G this presumes a homogeneous degree $d(v)$ for the nodes $v \in V$. A graph with this property is referred to as a regular graph. If the configuration space is not regular, a random walk will visit nodes with a relatively higher degree more frequently. Clustered nodes in G of an above average degree are referred to as areas of high connectivity in the following.

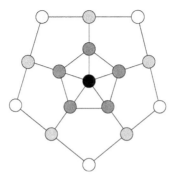

Fig. 1. Graph with heterogeneous node degree. The connectivity decreases with the distance from the center (black node).

An example is illustrated in Figure 1. The graph consists of 16 nodes of degrees 2, 3, 4 and 5 as indicated by different shades of gray. The six nodes of the inner pentagon form an area of high connectivity $(d(v) \geq 4)$. Once a random walk enters this region, it is more likely to remain in the region than it is to return to the perimeter. Note that this result holds only for undirected graphs, where symmetry is enforced. Clearly, variable $d(v)$ can bias any statistics based on a random walk. But an influence on stochastic local search may also be conjectured, motivated by the following metaphor.

Consider a sailor navigating in accordance with the direction and strength of the wind. After a while he recognizes that he is not going to reach the destination although direction and speed have been kept as calculated in advance. This is because the sailor is unaware of the tidal current offsetting his boat. The actual displacement depends on how well the tidal current matches the sailors course. Analogously, a local search algorithm navigates in accordance with the signal received from the fitness landscape. However, the underlying distribution of $d(v)$ may counteract its endeavor, and guide search to particular regions of the fitness landscape. We refer to this force as drift. The influence of drift on the outcome of local search depends on how well the connectivity of the landscape matches the overlying objective function.

In case of a regular graph $\hat{G}$ no drift exists. We conjecture this to be an important prerequisite for measuring the autocorrelation of a landscape $\hat{G}$ as described above. This is commonly expressed by the assumption of a statistically isotropic landscape. However, as shown in the next section, the configuration space of combinatorial optimization problems are not always regular.

3 A Model of the JSP Search Space

3.1 Configuration Space of the JSP

The general job-shop scheduling problem (JSP) is a well-known and widely-studied combinatorial optimization problem. The objective is to schedule a number of jobs on a number of machines such that the maximal completion time of the jobs involved is minimized [4].

In the disjunctive graph representation of solutions to the JSP, there exists an undirected disjunctive edge between each pair of distinct operations on a machine [1]. Selecting a particular orientation (i.e., precedence relation) for each disjunctive edge leads to a sequence of operations for each machine. The aggregate of machine sequences may lead to cyclic dependencies. Selections resulting in acyclic dependencies represent feasible schedules, with the length of the critical path corresponding to the schedule makespan. Fig. 2 shows the disjunctive graph for a JSP instance with $n = 3$ jobs (operations 1 and 2, operations 3 and 4, and operations 5 and 6 belong to the same job, respectively) and $m = 2$ machines. Notice that two jobs have an identical machine routing in this example.

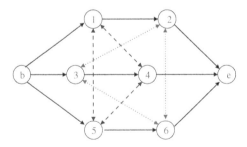

Fig. 2. Disjunctive graph representation for a 3-job 2-machine problem.

The selection of disjunctive arcs can be represented by a string of six bits where the first three bits correspond to precedence relations between operations on the first machine, and the last three bits corresponds to precedence relations between operations on the second machine. The total of $2^6 = 64$ combinations of precedence relations among operations can be represented by the 6-dimensional hypercube shown in Fig. 3(a). Here, neighboring nodes differ by one bit, corresponding to a difference in the orientation of exactly one precedence relation between a pair of consecutive operations on a machine.

The resulting neighborhood is known as the adjacent-swap (AS) neighborhood [2]. Moves under the AS neighborhood yield the smallest possible change to a schedule. Such a change is always reversible and for this reason the AS neighborhood is symmetric. Consequently, the configuration space G induced by the AS neighborhood can be viewed as an undirected graph.

In the disjunctive graph representation non-Hamiltonian selections are cyclic and therefore do not represent feasible operation sequences. Fig. 3(b) shows two sub-cubes of the hypercube where such infeasible solutions are located. The elimination of all one-machine cycles leads to 36 nodes in the hypercube of feasible solutions, cf. Fig. 3(c).

Valid sequences on individual machines can lead to cycles running over more than one machine. As a consequence, another 14 nodes can be eliminated from the hypercube, as shown in Fig. 3(d). The remaining 22 nodes correspond to the set of feasible schedules to the problem instance, cf. Fig. 3(e). The edges connecting two nodes express an AS move. One immediately recognizes a variable node degree $d(v)$. For instance, 100001 has a node degree of 4 while 000000 has node degree 2. Thus, the probability of visiting nodes in a random walk is not uniformly distributed, i.e., visiting 100001 is more likely than visiting 000000. Figure 3(f) verifies that the configuration space of the JSP is not regular. As with Figure 1, the node degree is depicted in gray shade.

3.2   Landscape of the JSP

In order to produce a fitness landscape for a problem instance, the processing times must also be taken into consideration. Operation processing times are introduced in the disjunctive graph model as weights for the arcs starting from the respective operation. For simplicity we assume identical processing times (say one unit of time) for the six operations in our example. This does not alter the feasibility of solutions and makes the precedence relations among operations more explicit. Due to this simplification, the makespan of a solution is given by the number of arcs that belong to the critical path of the corresponding graph representation. Notice that arcs starting from the initial dummy operation in the disjunctive graph carry zero weights and are therefore not counted.

The effect of identical processing times becomes obvious when representing schedules as Gantt charts. Since two of the three jobs have an identical machine routing, 11 qualitatively different "shapes" of Gantt charts are produced by the 22 solution nodes. As shown in Table 1 the makespan varies between 3 and 6 time units. The last column of the table gives the number of neighboring schedules (in terms of the AS neighborhood). In this example there is a correlation between the node degree of a solution and its makespan.

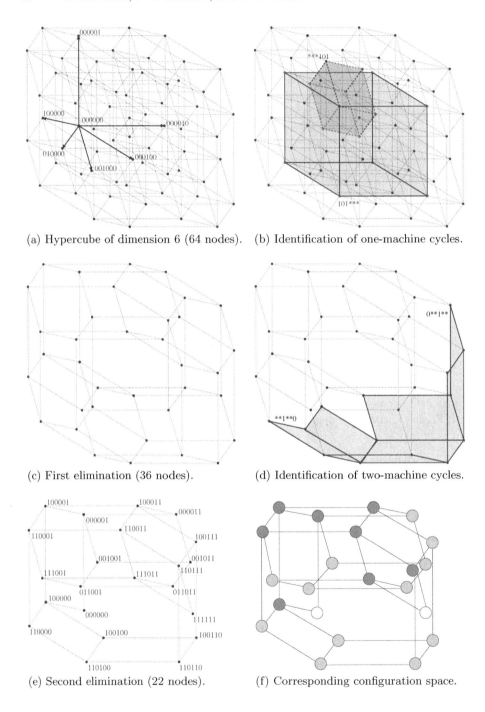

(a) Hypercube of dimension 6 (64 nodes).   (b) Identification of one-machine cycles.

(c) First elimination (36 nodes).   (d) Identification of two-machine cycles.

(e) Second elimination (22 nodes).   (f) Corresponding configuration space.

Fig. 3. Construction of the configuration space of the JSP instance.

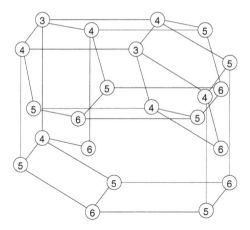

Fig. 4. Landscape of the JSP instance

The entire fitness landscape of the problem is shown in Fig. 4. The nodes are labeled with the corresponding makespan. In this small example every downhill walk terminates at one of the two global optima of 3 time units, i.e., the landscape is pseudo uni-modal. This example provides evidence of the existence of areas of high connectivity in the configuration space of the JSP. In such a landscape a random walk will proceed towards areas in the search space of higher connectivity. Since these areas show a better mean fitness, even a random walk should improve the solution quality to some extent.

4   Experimental Investigation

We now consider AS neighborhood connectivity in a real JSP, Fisher and Thompson's infamous 10×10 instance known as ft10 [6]. Our analysis is based on a set of 10,000 random semi-active solutions generated using the procedure described in [15]. A scatter-plot of solution makespan versus the number of AS neighbors $(d(v))$ is shown in Figure 5. The results clearly indicate that better solutions are located in regions of the configuration space with higher connectivity. Note that the maximal number of AS neighbors for this instance is 90. An analysis of the set of 13,120 optimal solutions to ft10 further reinforces this result, as $d(v)$ ranges between 88 and 89 for these solutions.

5   Conclusions and Future Research

We have shown that the fitness landscape of the JSP is highly irregular, in contrast to many other well-known combinatorial optimization problems. Irregularity leads to the potential for drift, in which a random walk is biased toward regions of the search space with higher degrees of connectivity. In the

Table 1. List of schedules belonging to the configuration space

Shape no.	Gantt chart	bitstrings	makespan	node degree
1		110011 100001	3	4
2		110001 100011	4	4
3		110111 100000	4	4
4		111011 000001	4	4
5		110110 100100	5	3
6		110000 100111	5	3
7		111001 000011	5	3
8		011011 001001	5	3
9		110100 100110	6	3
10		011001 001011	6	3
11		111111 000000	6	2

JSP, high-quality solutions generally possess high degrees of connectivity. As a consequence, we conjecture that such irregularity should aid both random walks and stochastic local search algorithms for the JSP, guiding search toward regions of the fitness landscape containing high-quality solutions.

Further research will address the relationship between efficient neighborhood moves defined in terms of the disjunctive graph model, and the probability of performing these moves in the face of an irregular landscape. The control of local search may be improved by considering the varying transition probabilities of an irregular landscape.

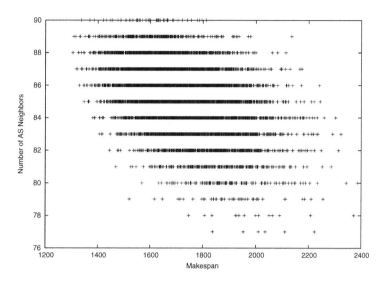

Fig. 5. Scatter-plot of makespan versus number of AS neighbors for 10,000 random semi-active solutions to ft10.

References

1. J. Adams, E. Balas, and D. Zawack. The shifting bottleneck procedure for job shop scheduling. *Management Science*, 34:391–401, 1988.
2. E.J. Anderson, C.A. Glass, and C.N. Potts. Applications in machine scheduling. In E.A. Aarts and J.K. Lenstra, editors, *Local Search in Combinatorial Optimization*, pages 361–414. John Wiley & Sons, 1997.
3. E. Balas and A. Vazacopoulos. Guided local search with the shifting bottleneck for job shop scheduling. *Management Science*, 44:262–275, 1998.
4. J. Błażewicz, W. Domschke, and E. Pesch. The job shop scheduling problem: Conventional and new solution techniques. *European Journal of Operational Research*, 93:1–33, 1996.
5. C. Smutnicki E. Nowicki. Some new ideas in ts for job-shop scheduling. In C. Rego and B. Alidaee, editors, *Adaptive Memory and Evolution. Tabu Search and Scatter Search*. Kluwer, 2003.
6. H. Fisher and G.L. Thompson. Probabilistic learning combinations of local job-shop scheduling rules. In *Industrial Scheduling*, pages 225–251. Prentice-Hall, 1963.
7. W. Hordijk. A measure of landscapes. *Evolutionary Computation*, 4:335–360, 1997.
8. A.S. Jain and S. Meeran. Deterministic job-shop scheduling: Past, present and future. *European Journal of Operational Research*, 113:390–434, 1999.
9. T. Jones. *Evolutionary Algorithms, fitness landscapes and search*. PhD thesis, University of New Mexico, Albuquerque, NM, 1995.
10. T. Jones and S. Forrest. Fitness distance correlation as a measure of problem difficulty for genetic algorithms. In *Proc of the 6th Int. Conf. on Genetic Algorithms*, pages 184–192. Morgan Kaufmann Publishers, 1995.
11. S.A. Kauffman. Adaptation on rugged fitness landscapes. In D. Stein, editor, *Lectures in the sciences of complexity*, pages 527–618. Addison-Wesley, 1989.

12. S. Kirkpatrick and G. Toulouse. Configuration space analysis for traveling salesman problems. *Journal de Physique*, 46(1277-1292), 1985.
13. M. Kolonko. Some new results on simulated annealing applied to the job shop scheduling problem. *European Journal of Operational Research*, 113:123–136, 1999.
14. P.J.M. Van Laarhoven, E.H.L. Aarts, and J.K. Lenstra. Job shop scheduling by simulated annealing. *Operations Research*, 40:113–125, 1992.
15. D.C. Mattfeld. *Evolutionary Search and the Job Shop*. Physica-Verlag, 1996.
16. D.C. Mattfeld, C. Bierwirth, and H. Kopfer. A search space analysis of the job shop scheduling. *Annals of Operations Research*, 86:441–453, 1999.
17. P. Merz and B. Freisleben. Fitness landscape analysis and memetic algorithms for the quadratic assignment problem. *IEEE Transactions on Evolutionary Computation*, 4:337–352, 2000.
18. E. Nowicki and S. Smutnicki. A fast tabu search algorithm for the job shop problem. *Management Science*, 42:797–813, 1996.
19. Colin R. Reeves. Landscapes, operators and heuristic search. *Annals of Operational Research*, 86:473–490, 1999.
20. Colin R. Reeves. Experiments with tunable fitness landscapes. In M. Schoenauer, K. Deb, G. Rudolph, X. Yao, E. Lutton, J.J. Merelo, and H.-P. Schwefel, editors, *Parallel Problem Solving from Nature - PPSN VI*, pages 139–148, 2000.
21. P.F. Stadler. Fitness landscapes. In M. Lässig and A. Valleriani, editors, *Biological and Statistical Physics*, pages 187–202. Springer Verlag, 2002.
22. P.F. Stadler and W. Schnabl. The landscape of the traveling salesman problem. *Physics Letters A*, 161:337–344, 1992.
23. É.D. Taillard. Comparison of iterative searches for the quadratic assignment problem. *Location Science*, 3:87–105, 1995.
24. J.-P. Watson, L.D. Whitley, and A.E. Howe. A dynamic model of tabu search for the job-shop scheduling problem. In G. Kendall, E. Burke, and S. Petrovic, editors, *Proc. of the 1st Multidisciplinary Int. Conf. on Scheduling: Theory and Applications*, pages 320–336. University of Nottingham, 2003.
25. E. Weinberger. Correlated and uncorrelated fitness landscapes and how to tell the difference. *Biol. Cybernetics*, 63:325–336, 1990.

A Comparison of Adaptive Operator Scheduling Methods on the Traveling Salesman Problem

Wouter Boomsma

Department of Computer Science,
University of Aarhus, Denmark
wb@daimi.au.dk

Abstract. The implementation of an evolutionary algorithm necessarily involves the selection of an appropriate set of genetic operators. For many real-world problem domains, an increasing number of such operators is available. The usefulness of these operators varies for different problem instances and can change during the course of the evolutionary process. This motivates the use of adaptive operator scheduling (AOS) to automate the selection of efficient operators. However, little research has been done on the question of which scheduling method to use. This paper compares different operator scheduling methods on the Traveling Salesman Problem. Several new AOS techniques are introduced and comparisons are made to two non-adaptive alternatives.

The results show that most of the introduced algorithms perform as well as Davis' algorithm while being significantly less cumbersome to implement. Overall, the use of AOS is shown to give significant performance improvements – both in quality of result and convergence time.

1 Introduction

Genetic operators are the algorithmic core of evolutionary computation. The quality of these operators is essential for the performance of the evolutionary algorithms (EAs), and consequently, much research is done in this field. Particularly for combinatorial problems, domain knowledge is often essential to good performance. Many heuristic operators have been described, and their number continues to grow. Given their heuristic nature, the applicability of these operators often varies across different problems in the domain. For instance, certain operators might fail to scale up gracefully, becoming computationally infeasible for larger problem instances. Even though the literature might provide comparisons between operators on certain problem instances, the evolutionary programmer is left with a difficult choice of operator selection when designing an EA for a new problem.

An elaborate manual comparison of all operators would provide an assessment of the quality of the operators, but does not exploit the fact that the usefulness of the operators often changes during the course of a single run. Furthermore, dependencies between operators can exist and through interaction multiple operators might provide better results than when applied alone.

J. Gottlieb and G.R. Raidl (Eds.): EvoCOP 2004, LNCS 3004, pp. 31–40, 2004.

In a previous study [1], I investigated whether adaptive operator scheduling (AOS) can provide a solution to this problem. Experiments were done on instances of the symmetrical Traveling Salesman Problem (TSP), a well known NP-hard combinatorial problem for which a multitude of operators exist. It can be defined as the search for a minimal Hamiltonian cycle in a complete graph, and can be understood as the problem of visiting n cities (and returning to the first), using the path of smallest total distance. The main concern in the original investigation was that a large number of operators might slow down the optimisation process compared to an algorithm using the optimal choice of operators. It was shown that this concern was unfounded: the algorithm using AOS converged as fast as the best combination of mutation and crossover operators with equally good results. The AOS scheme used in these initial experiments (Davis) was however rather cumbersome to implement. In this paper several alternative methods of AOS are presented and compared on a selection of TSP benchmark problems. It is shown that the presence of multiple operators significantly improves performance. Furthermore, results indicate that the alternative AOS methods perform as well as Davis' method, and that simpler methods can thus be used without a decrease in performance.

2   Adaptive Operator Scheduling

Angeline [2] has categorised adaptive evolutionary computation based on two criteria: (1) the level at which adaptation is applied and (2) the nature of the update rules. The level of adaptation specifies at which level the adapted parameters reside. For population-level adaptation, the parameters exist at the population level, and thus apply to all individuals in the population. Likewise, individual-level adaptation works on parameters local to each individual and component-level adaptation has parameters for each component in the genotype of an individual. Update rules are classified as being absolute, or empirical This last class also goes by the name of self-adaptation, which is the term that will be used throughout this paper. Algorithms with absolute update-rules use fixed guidelines to update parameter settings based on the current state of the population. Self-adaptive schemes use the selection pressure already present in the evolutionary process to evolve better parameter settings, the underlying assumption being that good individuals often have a good parameter setting.

The AOS algorithms described below represent different classes of Angeline's classification. Focus was on designing AOS algorithms that are easily implemented, which is vital if AOS is to replace the manual comparison of individual operators. The end of each description contains a list of the parameter settings that were used for the algorithm. These settings were manually tuned based on preliminary experiments.

2.1   The Operator Scheduling Algorithm by Davis

Davis' algorithm from 1989 [3] represents one of the first efforts at operator adaptation in EAs. It uses population-level adaptation and absolute update rules on a steady state EA. More specifically, a global set of operator probabilities is adapted based on the performance of the operators in the last generations (adaptation window). The performance of operators is measured by the quality of the individuals they produce. Newly created individuals are rewarded if their fitness surpasses the fitness of all other individuals in the population. The size of this reward is determined by the amount of the improvement. Furthermore, a certain percentage of the reward is recursively passed on to the individual's ancestors (to a certain maximum depth M). This reward strategy is motivated by the fact that a series of suboptimal solutions is often necessary in order to reach a better solution, and that corresponding operators should thus be rewarded.

With certain intervals, the rewards are used to update the probability setting. For each operator i:

$$p'_i = (1 - S) * p_i + S * \frac{\text{reward}_i}{\text{totalReward}}, \tag{1}$$

where p'_i is the new probability for operator i and S is the Shift factor, determining the influence that the current update should have on the total probability setting.

In the present study a slightly modified version of Davis' algorithm is used. Lower bounds are set on the probabilities to avoid extinction of operators, and in adaption phases where no improvement is made the probabilities are shifted slightly toward their initial positions. Furthermore, the steady state evolutionary approach is replaced by a generational elitist EA. Preliminary experiments showed that these modifications gave better results and faster convergence (results not shown).

Parameter settings: Window of adaptation (W): 100 individuals, Interval of adaptation (I): 20–50 evaluations, Shift percentage (S): 15%, Percentage of reward to pass back (P): 90%, Number of generations to pass back (M): 10.

2.2   The ADOPP Algorithm

Julstrom's Adaptive Operator Probabilities algorithm (ADOPP) [4] from 1995 is very similar to Davis' approach. The main difference between the two is the way in which ancestral information is represented. While Davis provides each individual with links to their parents, Julstrom explicitly provides each individual with a tree specifying which operators were used to create its ancestors. Davis' approach is more effective since the tree structure is implicitly present in the individuals and does not have to be copied every time new individuals are created. However, Davis' algorithm has some disadvantages which Julstrom avoids. Since the ancestral information in Davis' algorithm is stored in the individuals, information is lost when individuals die. The depth of the implicit trees are therefore somewhat unreliable and, in general, Davis' algorithm is only able to maintain a moderate amount of ancestral information. Furthermore, the individuals in Davis' algorithm require a great deal of bookkeeping to sustain pointers only to living individuals, an inconvenience which is avoided in ADOPP.

Another difference between the two approaches is that Davis rewards individuals that improve the overall best individual in the population, whereas Julstrom only requires individuals to exceed the median individual.

When converting operator rewards to a new probability setting, ADOPP uses a greedy variant of the update rule by Davis (corresponding to S=100%). For each operator i:

$$p'_i = \frac{\text{reward}_i}{\text{totalReward}}, \tag{2}$$

where p'_i is the new probability for operator i.

Parameter settings: Window of adaptation (QLEN): 100 individuals, Interval of adaptation (I): 1 generation, Percentage of reward to pass back (DECAY): 80%, Height of trees (DEPTH): 4.

2.3   Adaptation Using Subpopulations (Subpop)

This approach was inspired by the work by Schlierkamp-Voosen and Mühlenbein [5] on competing subpopulations, in which different subpopulations represent different operator strategies. This idea can be used as an AOS method by letting each subpopulation represent the use of a single operator. The EA thus maintains a number of subpopulations equal to the number of operators. For each of these populations, only one operator can be applied to the individuals currently residing there. During the course of a run, the relative sizes of the subpopulations are altered depending on their fitness. The fitness of a subpopulation is based on the fitness of its best individuals over the last 10 generations (as proposed by Schlierkamp-Voosen and Mühlenbein [5]).

In each adaptation phase the best subpopulation is rewarded with an increase in size, and receives a donation of individuals from all other subpopulations. To avoid the extinction of operators, only subpopulations with a size above some fixed lower bound are forced to make this donation. Large subpopulations create more offspring and thus have a larger probability of improving the global best fitness. This results in a strong bias towards larger subpopulations, making it difficult for smaller populations to compete. To counter this effect, a random migration scheme is used: At certain intervals some of the best individuals from the best population migrate to a randomly selected population.

Subpop is a population-level adaptation algorithm using absolute update rules. The parameters adapted are the same as for Davis and ADOPP (global operator probabilities).

Parameter settings: Evaluation interval (E): 4 generations, Shift amount (SA): 10%, Migration interval (M): 4 generations, Migration amount (MA): 5 individuals, Interval of adaptation (I): 1 generation.

2.4   Self-Adaptive Operator Scheduling (SIDEA, SPDEA)

The algorithms in this section are novel methods of operator adaptation partly inspired by the work of Spears in 1995 [6]. In Spears' paper, scheduling is done between two crossover operators by adding a single bit to the genotype indicating the preferred operator. This bit is used either locally (at the individual-level) or globally (at the population-level) to determine which operator is to be applied. When used locally, the choice of operator is made based on the bit-setting of the parent(s) involved. When used globally, the average bit-setting of the whole population is used as a measure of quality to base this decision on.

I generalised Spears' method by expanding the single bit to an array of n probabilities, each denoting the probability that a certain operator is applied. Unlike Spear's approach, the representation of the scheduling information is not compatible with that of the problem representation and therefore cannot be modified automatically as part of the genotype. It was therefore necessary to design a specialised variation operator for this task.

The problem of finding the optimal probability setting for n operators is a numerical optimisation problem so in principle, any variation operator from this domain would suffice. However, given the fact that the probabilities must all be positive and sum to one, only a small subset of the n-dimensional search space constitutes legal solutions. If an arbitrary EA variation operator is used it would be necessary to apply some repair scheme after each operation to ensure feasibility. To avoid this I used an adapted version of the variation operator used in Differential Evolution (DE) [7], which produces solutions that are only rarely illegal. The DE variation operator uses 3 individuals to create an offspring by applying a mutation step followed by a crossover step. During the mutation step, a new genotype x' is created by:

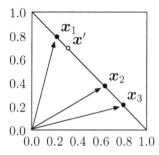

Fig. 1. The DE mutation operator for two dimensions. Three random individuals x_1, x_2, x_3 are chosen and a temporary individual $x^{\square}$ is created by taking the vectorial difference between x_3 and x_2, multiplying it by F, and adding it to x_1.

$$x' = x_1 + F * (x_3 - x_2) . \qquad (3)$$

The crossover step subsequently creates a new individual by combining components from x' with components from the original individual. For our purpose, however, the mutation step alone has exactly the properties we need: Selecting three random points on a hyperplane and adding the vectorial difference between two of them to the third results in a new position on the plane. In other words, given three legal probability settings it will produce a new one with probabilities that sum to 100%. An example of this behaviour for two dimensions is given in Fig. 1. Equation (3) of course gives no guarantees that the entries of the created vector have values between zero and one. This is however easily fixed by forcing values that lie outside the domain back to the nearest boundary. The mutation operator is used whenever new offspring is to be created.

As with Spears' algorithm, two different levels of adaptation exist. If individual-level adaptation is used, the probability setting of the child is used to choose an operator. If adaptation works at the population-level, the average of all probability-settings in the population is used to select an operator. The two versions of the algorithm will be referred to as the SIDEA and the SPDEA respectively.

Parameter setting: F-value for the DE mutation: 0.5.

2.5   Operator Scheduling Inspired by Evolution Strategies (SIESA)

Self-adaptation has been applied in the Evolution Strategies (ES) community since 1977 [8]. Here it was used to adapt the behaviour of the mutation operator during the course of a run. The mutation operator consists of an addition of normally distributed values $z_i \sim \mathbf{N}(0, \sigma_i'^2)$ to the components in the genotype. The variances $\sigma_i'^2$ are adapted so that the effect of mutation is different for different individuals and for different components of the genotype. Again, under the assumption that individuals of high fitness often have good parameter settings, the algorithm can find the good variance setting using only the selection pressure already present in the algorithm.

Taking self-adaptation one step further, we now use the ES mutation step as an alternative to the DE mutation operator described in the previous section. This means that the evolutionary algorithm will simultaneously optimise the problem at hand, an operator probability setting, and a setting determining the optimal variance for the mutation of this probability setting.

Since the search space defined by the operator probability settings is highly interdependent, it is meaningless to adapt each variance σ_i^2 at component-level. Instead one value σ is associated with each individual. The necessary mutation equation from the ES literature [9] is

$$\sigma' = \sigma * \exp(s_0), \tag{4}$$

where $s_0 \sim \mathbf{N}\left(0, \tau_0^2\right)$ and $\tau_0^2 = \frac{1}{n}$. Since this mutation operator does not have the convenient properties of the DE operator, the probability-setting has to be normalised after each mutation. It is difficult to tell exactly how big the impact of this normalisation is compared to the effect of the mutation itself.

Parameter setting: τ_0^2: $\frac{1}{n}$ (ES default value [9]).

3   Operators

Table 1 lists the 18 operators used in the experiments. They all operate on a path representation of the TSP. The selection of operators was inspired by Larrañaga's survey paper from 1999 [10]. The list was extended by two operators of recent date that have been shown to have good performance: The Edge Assembly Crossover [11] and the Inver-over operator [12]. For complete references the reader is referred to the paper covering my previous experimentation on AOS for the TSP [1].

Table 1. The Operators

Displacement Mutation (DM)	Order Based Crossover (OX2)
Exchange Mutation (EM)	Position Based Crossover (POS)
Insertion Mutation (ISM)	Heuristic Crossover (HEU)
Simple Inversion Mutation (SIM)	Edge Recombination Crossover (ER)
Inversion Mutation (IVM)	Maximal Preservative Crossover (MPX)
Scramble Mutation (SM)	Voting Recombination Crossover (VR)
Partially mapped Crossover (PMX)	Alternating Position Crossover (AP)
Cycle Crossover (CX)	Inver-over operator (IO)
Order Crossover (OX1)	Edge Assembly Crossover (EAX)

4   Experiments and Results

Initial experiments showed that the algorithms benefitted from larger populations as the problem instances grew, confirming the findings by Nagata and Kobayashi [11]. The large populations are however only partly replaced in each generation. The ADOPP variant uses a steady state approach, generating only one new individual in each generation (in agreement with Julstrom's original paper [4]). The other algorithms were run with elite sizes of 75% – 90% of the population, which proved to give best results in initial experiments. With these extensive elite sizes the large populations function mainly as libraries of genetic material which maintain a certain diversity in the population. The population sizes and number of iterations used are listed in Table 2.

Table 2. Population sizes and iterations used for the different problems

	gr48	brg180	pcb442	nrw1379	pr2392	pcb3038
Population Size	500	600	2000	4000	4500	5000
Iterations	60,000	350,000	500,000	1,200,000	1,200,000	1,700,000

When implementing an AOS algorithm, one has the choice of including all operators in one pool or dividing mutation and crossover operators in two separate pools. For Davis, ADOPP, SIDEA and SPDEA, both variants were implemented and included in the test set. The versions using separate pools are labelled with the suffix _S. Two non-adaptive algorithms were included in order to evaluate the absolute value of AOS. The Uniform algorithm includes all operators but uses equal probabilities for the application of them. The EAX_only uses only the EAX operator.

For all algorithms, 100 runs were done on 6 different TSP instances, all taken from the TSP benchmark problem collection TSPLIB [13]. The sizes of the problems ranged from 48 to 3038 cities. Table 3 presents the average results of these trials.

To give an impression of convergence speed Fig. 2 shows the average best fitness over time for the brg180 problem. The general pattern of this figure is recurrent across the range of tested problems: The EAX_only algorithm performs

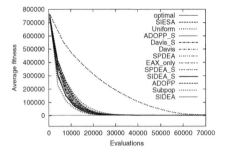
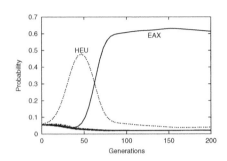

Fig. 2. Convergence graph for all algorithms on the brg180 problem

Fig. 3. Operator probability distribution during an execution of SIDEA on brg180

significantly worse than all others, the Uniform algorithm performs almost as good as the adaptive algorithms, and among the adaptive algorithms, Subpop tends to have the fastest convergence. Figure 3 shows the operator probabilities over time for the SIDEA algorithm applied on the brg180 problem. This figure is also representative for all cases: Across both problem instances and AOS methods all algorithms consistently prioritise the EAX operator in the final phase. In the initial phases the greedy nature of the other operators (typically HEU) is most effective to gain fast fitness improvements. This clearly improves performance compared to the algorithm applying the EAX operator exclusively.

Based on the data in Table 3, it is not possible to single out one algorithm as being the overall best. However, a certain categorisation of the algorithms can be made. Across the range of problems, the Davis(_S), SIDEA(_S), SPDEA(_S) and Subpop algorithms all performed about equally well. The data responsible for the mean values in the table were found to be approximately normally distributed. Standard deviations and confidence intervals for the results were computed. Although there were non-overlapping confidence intervals between results in Table 3, and differences thus can be seen as being statistically significant, it varies from problem to problem which algorithm performed best.

The ADOPP and SIESA algorithms performed slightly worse than this first class. For ADOPP's case this might be ascribed to the algorithm being more greedy than, for instance, the Davis algorithm, and therefore might overcompensate bad operators when they by chance improve the overall solution. On the other hand, it seems that the SIESA adapts too slowly. This is not entirely surprising, since it is using self-adaptation at two levels. Another reason could be the somewhat disruptive renormalisation after each operation.

Perhaps the most striking results in Table 3 concern the two non-adaptive algorithms. The EAX_only algorithm performs significantly worse than any of the other algorithms. This suggests that the EAX operator is first and foremost a fine-tuning operator and should be used along with some more exploration-oriented operators. The Uniform algorithm, on the other hand, performs remarkably well. Especially the convergence graphs show surprisingly fast convergence. It should however be noted that the convergence graphs show the population's best fitness over time and that no information about the general state of the population can be derived from these graphs. The apparent conflict between the Uniform algorithm's good convergence speed and the somewhat worse end-results of this algorithm indicates that even though the algorithm is able to sustain a good elite of individuals, it does not have sufficient diversity in its best individuals to find optimal solutions. The smaller number of good individuals is naturally explained by the fact that in the end phase, the EAX operator is applied with only a fraction of the probability that it receives by the adaptive algorithms.

Generally, the adaptive operator scheduling methods did not present a significant computational burden for the EAs involved. The different algorithms had similar execution times, except for the ADOPP variants, which were somewhat slower due to their steady state design.

Table 3. Average results for 100 runs

	gr48	brg180	pcb442	nrw1379	pr2392	pcb3038
Davis_S	5046.00	1950.30	50778.21	56689.46	378067.16	137758.94
Davis	5046.36	1950.00	50781.23	56685.04	378055.16	137752.69
SIDEA_S	5047.77	1950.20	50778.14	56699.82	378084.82	137779.77
SIDEA	5046.52	1950.30	50778.07	56715.85	378336.45	137793.97
SPDEA_S	5046.33	1950.40	50778.21	56698.31	378097.53	137781.66
SPDEA	5046.00	1950.30	50778.07	56691.05	378074.36	137755.16
SIESA	5047.21	1952.60	50781.83	56695.51	378098.00	137765.02
Subpop	5046.40	1950.60	50778.14	56682.53	378057.28	137743.03
Uniform	5048.34	1952.80	50778.91	56826.32	379570.91	137956.90
ADOPP_S	5046.66	1951.80	50778.56	56702.35	378100.71	137779.00
ADOPP	5046.66	1951.70	50778.35	56689.42	378079.92	137745.00
EAX_only	5046.00	1951.30	59558.96	128619.73	2823286.85	677724.28
Optimal	5046.00	1950.00	50778.00	56638.00	378032.00	137694.00

5 Discussion

Overall, the results show that performance is significantly improved when a multitude of operators are used. Even when one operator (e.g. EAX) is suspected to outperform all others, it might not be optimal throughout the whole run, and adaptive operator scheduling can exploit this fact to increase performance.

Based on the comparison of operator scheduling algorithms in this study, it is not possible to single out one of them as being the best. Many of the different methods performed equally well, and thus indicate that the selection of an operator scheduling method may be based on considerations as implementation efficiency or personal taste. The Davis algorithm requires significant bookkeeping and was found to be somewhat elaborate to implement, while especially the SIDEA, SPDEA, SIESA and Subpop variants were implemented fairly easily. Also the fact that the self-adaptive variants have only few parameters to tune might be a reason to favour these over the others. As an extreme case, even a uniform selection between operators seems to provide reasonable results at minimal implementation costs. Naturally, experiments should be carried out on other problem domains to determine the value of operator scheduling in general and establish whether the above conclusions hold for other domains.

References

1. Boomsma, W.: Using adaptive operator scheduling on problem domains with an operator manifold: Applications to the travelling salesman problem. In: Proceedings of the 2003 Congress on Evolutionary Computation (CEC2003). Volume 2. (2003) 1274–1279
2. Angeline, P.J.: Adaptive and self-adaptive evolutionary computations. In Palaniswami, M., Attikiouzel, Y., eds.: Computational Intelligence: A Dynamic Systems Perspective. IEEE Press (1995) 152–163
3. Davis, L.: Adapting operator probabilities in genetic algorithms. In Schaffer, J.D., ed.: Proceedings of the Third International Conference on Genetic Algorithms, San Mateo, CA, Morgan Kaufman (1989)
4. Julstrom, B.A.: What have you done for me lately? Adapting operator probabilities in a steady-state genetic algorithm. In Eshelman, L., ed.: Proceedings of the Sixth International Conference on Genetic Algorithms, San Francisco, CA, Morgan Kaufmann (1995) 81–87
5. Schlierkamp-Voosen, D., Mühlenbein, H.: Strategy adaptation by competing subpopulations. In Davidor, Y., Schwefel, H.P., Männer, R., eds.: Parallel Problem Solving from Nature – PPSN III, Berlin, Springer (1994) 199–208
6. Spears, W.M.: Adapting crossover in evolutionary algorithms. In McDonnell, J.R., Reynolds, R.G., Fogel, D.B., eds.: Proc. of the Fourth Annual Conference on Evolutionary Programming, Cambridge, MA, MIT Press (1995) 367–384
7. Storn, R., Price, K.: Differential evolution - a simple and efficient heuristic for global optimisation over continuous spaces. Journal of Global Optimization **11** (1997) 341–359
8. Bäck, T., Hoffmeister, F., Schwefel, H.P.: A survey of evolution strategies. In Belew, R., Booker, L., eds.: Proceedings of the Fourth International Conference on Genetic Algorithms, San Mateo, CA, Morgan Kaufman (1991) 2–9
9. Bäck, T.: Evolution strategies: An alternative evolutionary algorithm. In Alliot, J., Lutton, E., Ronald, E., Schoenhauer, M., Snyers, D., eds.: Artificial Evolution, Springer (1996) 3–20
10. Larrañaga, P., Kuijpers, C.M.H., Murga, R.H., Inza, I., Dizdarevic, S.: Genetic algorithms for the travelling salesman problem: A review of representations and operators. Artificial Intelligence Review **13** (1999) 129–170
11. Nagata, Y., Kobayashi, S.: Edge assembly crossover: A high-power genetic algorithm for the travelling salesman problem. In Bäck, T., ed.: Proceedings of the Seventh International Conference on Genetic Algorithms (ICGA97), San Francisco, CA, Morgan Kaufmann (1997)
12. Tao, G., Michalewicz, Z.: Inver-over operator for the TSP. In Eiben, A.E., Bäck, T., Schoenauer, M., Schwefel, H.P., eds.: Parallel Problem Solving from Nature – PPSN V, Berlin, Springer (1998) 803–812
13. Reinelt, G.: TSPLIB — a traveling salesman problem library. ORSA Journal on Computing **3** (1991) 376–384

AntPacking – An Ant Colony Optimization Approach for the One-Dimensional Bin Packing Problem

Boris Brugger[1], Karl F. Doerner[1], Richard F. Hartl[1], and Marc Reimann[1,2]

[1] Department of Management Science, University of Vienna,
Bruenner Strasse 72, A-1210 Vienna, Austria
{Karl.Doerner, Richard.Hartl}@univie.ac.at
[2] Institute for Operations Research, Swiss Federal Institute of Technology Zurich,
Clausiusstrasse 47, CH-8092 Zurich, Switzerland
Marc.Reimann@ifor.math.ethz.ch

Abstract. This paper deals with the one-dimensional bin packing problem and presents a metaheuristic solution approach based on Ant Colony Optimization. Some novel algorithm design features are proposed and the comprehensive computational study performed, shows both the contribution of using these features as well as the overall quality of the approach as compared to state of the art competing metaheuristics.

1 Introduction

Bin packing type problems occur in many practical situations such as transportation, multi-period capital budgeting problems, parallel machine scheduling problems and knapsack problems. Following Falkenauer (1996) the one-dimensional bin packing problem (BPP) can be stated as follows: "Given a finite set O of numbers and two constants C and N, is it possible to pack all the items into N bins - that is, does there exist a partition of O into N or less subsets, such that the sum of elements in any of the subsets does not exceed C?"

Garey and Johnson (1979) show that this problem belongs to the class of NP-hard problems. Hence, exact solution techniques are bound to work well for small to medium sized problem instances only and real world sized problems including up to thousands of items have to be solved heuristically.

In recent years, meta-heuristic approaches have been applied to the one-dimensional BPP at the forefront of which algorithms based on evolutionary concepts have to be mentioned. Falkenauer (1996) developed a Hybrid Grouping Genetic Algorithm (HGGA) which to date is the most effective mate-heuristic approach for the one-dimensional BPP. The main idea of this approach is to form offspring by choosing a set of bins from each parent solution, eliminating duplicate items and packing the remaining unpacked items according to the simple First Fit Decreasing (FFD) heuristic. Further, a local search based on a dominance result of Martello and Toth (1990) is used to improve the offspring both before and after the FFD repair. The main idea of this local search is to

J. Gottlieb and G.R. Raidl (Eds.): EvoCOP 2004, LNCS 3004, pp. 41–50, 2004.

exchange up to three items of a given bin by one or two larger and still unpacked items, such that the considered bin has an improved filling degree. Apart from this direct improvement in the utilization of the considered bin, an indirect improvement is achieved as smaller items, which are possibly easier to pack into the other bins, are left unassigned.

Another approach for the one-dimensional BPP proposed by Levine and Ducatelle (2004) is based on Ant Colony Optimization (ACO). More precisely, a Hybrid Max-Min Ant System (denoted by HACO) is used. The solution construction mechanism used in this approach is based on the FFD heuristic and pheromone decoding is based on the pairwise favorability of item lengths in the same bin. The local search employed to improve the ants' solutions is again based on the dominance result of Martello and Toth (1990). However, Levine and Ducatelle (2004) restrict the number of items to be replaced to two. The results show that the HACO approach finds comparable results to the HGGA.

In this paper we propose a different variant of an ACO algorithm for the one-dimensional BPP. In fact, we apply a modified Ant System (AS), where we employ the Ant Colony System (ACS) decision rule. Apart from that we use a different pheromone decoding, a different fitness function for the pheromone update and a preprocessing step to reduce the search space for the ants. Further we apply the local search during the solution construction each time a bin is closed.

We perform an extensive numerical analysis to analyze the design decisions and to show the excellent performance of the approach when compared to both the HGGA and the HACO.

The remainder of this paper is organized as follows. In the next section we present in detail our AntPacking approach. This will be followed by the results of our numerical analysis in Section 3. Finally, we conclude in section 4 with a summary of our work and an outlook on further research issues for solving the BPP with ACO.

2   The AntPacking Algorithm

Our AntPacking algorithm is based on ACS, which is a particular realization of the ACO metaheuristic first proposed by Dorigo and Di Caro (1999). Other ACO algorithms have been successfully applied to a variety of combinatorial optimization problems such as the TSP, the QAP, different variants of the VRP, the Graph Coloring Problem and different variants of machine scheduling problems. For an overview of the most successful applications we refer to Dorigo and Stuetzle (2004). A convergence proof for an ACO algorithm can be found in Gutjahr (2002).

The main idea of ACO is that a population of computational ants repeatedly builds and improves solutions to a given instance of a combinatorial optimization problem. From one generation to the next a joint memory is updated that guides the work of the successive populations. The memory update is based on the solution quality of the ants and is more or less biased towards the best solutions found. In the extreme cases of ACS and Max-Min Ant System only the overall best solution found up to a given generation is reinforced. Moreover, the reinforcement relates to the individual components of these solutions, such that e.g. in the TSP all arcs occurring in the best solution are reinforced.

Let us now turn to a description of the application of this metaphor to the one-dimensional BPP.

2.1   Pheromone Decoding

As stated in the introduction the pheromone decoding used in this paper is one of the contributions of our approach. While Levine and Ducatelle (2004) used a decoding based on the lengths of the items, which relates the length of an item to be added to the lengths of all items already in the bin, we use quite a different approach.

First, we follow Levine and Ducatelle (2004) and group items according to their lengths. However, we then relate the lengths of an item to be packed to the filling degree of the current bin. Thus, we disregard the information about which items are already in the bin. Rather, the only thing that is important is how much space is left in the bin. The pheromone matrix thus has one row for each item length and one column for each filling degree.[1] An entry in this matrix is thus denoted by $\tau_{l_i, F_{b,i}}$, where l_i is the length of item i and $F_{b,i}$ is the filling degree of a bin b before item i is to be filled in. By doing so we directly emphasize the importance to fill a bin as good as possible as opposed to emphasizing certain combinations of items that fill a bin as good as possible.

2.2   Solution Construction and Local Search

Each ant constructs a solution by sequentially filling bins until all the items have been packed. A new bin is opened only if no more item can be added to the current bin.

The decision about which item to add to the current bin is based on both a heuristic criterion and the pheromone information. The heuristic criterion used is based on the FFD rule. Thus a longer item is more likely to be chosen than a shorter one, i.e. $\eta_i = l_i$, where l_i is the length of item i. Clearly an item can only feasibly be added if its inclusion does not violate the capacity of the bin.

The decision rule is based on an explicit modelling of exploitation. More specifically, a parameter q_0 is added that determines the probability to exploit the best option available in a decision step. One advantage of this decision rule, besides the increased exploitation of the best solution, is the increase in speed, as the time consuming roulette wheel selection is only applied with a probability $1 - q_0$.

Summarizing, let Ω denote the set of unpacked items that can be added to the current bin b without violation of the capacity constraint, i.e. $i \in \Omega$, iff $F_{b,i} + l_i <= C$.

---

[1] Note, that item lengths and bin capacity are integer values.

Then, the decision rule can formally be written as

$$
P_i(F_{b,i}) = \begin{cases} 0 & \text{if } i \notin \Omega \\[2mm] q_0 + (1 - q_0) \dfrac{\eta_i^\beta \tau_{l_i, F_{b,i}}^\alpha}{\sum_{j \in \Omega} \eta_j^\beta \tau_{l_j, F_{b,j}}^\alpha} & \text{if } \eta_i^\beta \tau_{l_i, F_{b,i}}^\alpha = \max_{j \in \Omega} \eta_j^\beta \tau_{l_j, F_{b,j}}^\alpha \\[3mm] (1 - q_0) \dfrac{\eta_i^\beta \tau_{l_i, F_{b,i}}^\alpha}{\sum_{j \in \Omega} \eta_j^\beta \tau_{l_j, F_{b,j}}^\alpha} & \text{otherwise.} \end{cases} \tag{1}
$$

Once an ant has filled a bin and there are no items left that fit into the bin without violating its capacity a local search motivated by the FFD and the dominance results of Martello and Toth (1990) is evoked. One by one it tries to replace one item of the bin with an unpacked item if this leads to a better capacity utilization of the bin and does not violate the capacity constraint. The local search scans the packed items in the sequence of their packing, and the unpacked items according to their lengths. Thus, it will always add the longest feasible unpacked item. Further, a possible local search move will result in both, a better utilization of the current bins' capacity and a smaller unpacked item.

The local search stops if the bin is filled completely or there are no more improving moves possible. At this point the ant opens the next bin and repeats the construction and local search steps until all items are packed.

2.3   Fitness Function and Pheromone Update

Evolutionary algorithms and also ACO approaches need a fitness function that guides the learning and consequently the search. Classically in a minimization problem the fitness is just calculated as the inverse of the objective function value of a given solution. Hence, the trivial way to evaluate the quality of a solution in the BPP is just to take the inverse of the number of bins used. However, as pointed out by Falkenauer and Delchambre (1992) this leads to a very particular shape of the search landscape where an optimal solution is surrounded by a large number of local optima with one additional bin. Clearly such a landscape is not well suited for learning algorithms if there is no way to distinguish between these local optima.

Thus, Falkenauer and Delchambre (1992) propose a fitness function, where the maximization of the average bin utilization (which corresponds to minimizing the number of bins) is augmented by the consideration of the variance in bin utilization. This is achieved by adding a parameter z, which helps to discriminate between solutions with equally filled bins and ones in which some bins are rather full and others rather empty. The formal representation of this fitness function is given as follows,

$$
f(s) = \frac{\sum_{b=1}^{N(s)} \left(\frac{F_b}{C}\right)^z}{N(s)}, \tag{2}
$$

where $N(s)$ is the number of bins used in solution s, $f(s)$ is the fitness of solution s, C is the capacity of the bins and F_b is the total filling degree of bin b. Clearly, for $z = 1$ we get the trivial fitness equal to the inverse of the number of bins. On the other hand, as z increases a solution with larger differences in bin utilization gets more and more favorable as compared to a solution with equally filled bins. The intuition is that the former solution will be a better candidate for getting rid of some (rather empty) bin.

This fitness function was also used by Levine and Ducatelle (2004). However, in the context of an ACO algorithm, the fitness function is needed for two purposes. First, it is used to compare the ants' solutions with one another. Second, it is also used to reinforce the solution elements of the best solutions determined in the first step. Levine and Ducatelle (2004) use the above mentioned fitness function for both purposes.

In this paper, we use quite a different approach. First, we do not discriminate between the ants. Rather, as is typical for AS, all ants are allowed to update the memory. Second, the fitness evaluation used for this update is completely new. In fact, our intuition is to reinforce patterns in the form of well filled bins rather than complete solutions. Thus, we use a special form of the function above, which we will call local, binary fitness function.

Using this function, we look at bins locally by evaluating each bin of a given solution separately. In addition to that, we use an extreme case, where only full bins are reinforced, whereas partly filled bins do not receive any pheromone. The resulting fitness formulation for each bin is

$$f_{b(s)} = \begin{cases} 1 \text{ if } F_b = C \\ \\ 0 \text{ if } F_b < C \end{cases} \quad \forall b = 1...N(s) \quad (3)$$

Note, that the fitness is now calculated for each bin b separately. Note also that for each bin this function corresponds to a case where $z = \infty$.

Having described our definition of fitness for each bin, we can now turn to the pheromone update.

The pheromone update is based on the classical Ant System pheromone update. Thus, each ant is allowed to modify the memory. Moreover, all the solution elements are subject to evaporation. However, as pointed out above only completely filled bins will be reinforced, regardless of the quality of their associated complete solution. Formally this can be written as

$$\tau_{l_i, F_{b,i}} = (1 - \rho)\tau_{l_i, F_{b,i}} + \rho\Delta\tau_{l_i, F_{b,i}}, \quad (4)$$

where

$$\Delta\tau_{l_i, F_b} = \sum_s f_{b_i(s)} \quad \forall s \in S. \quad (5)$$

Here $b_i(s)$ denotes the bin item i is assigned to in solution s and S corresponds to the number of solutions generated in the current iteration, i.e. the number of ants. Finally, the pheromone matrix is initialized with a value τ_0, which is a parameter of the algorithm and determines the breadth of the search in the initial phase.

2.4   Preprocessing

Now that we have depicted the main components of our AntPacking algorithm, we will turn to a preprocessing procedure which we added to reduce the search effort for the ants. This preprocessing is based on the format of the problem instances. More precisely, in the benchmark instances bin capacity is set to $C = 150$ and the individual item lengths' are drawn uniformly from the integers bounded by $[20, 100]$.

This format of the problem data leads to two observations. First, a bin can never be filled completely with just one item. Second, if a bin can be filled completely with two items this filling will be part of an optimal solution. Thus, by searching for pairs of items that fill bins completely and eliminating these items from further consideration we can reduce the search space for the ants considerably. In fact, this is exactly what our preprocessing step does.

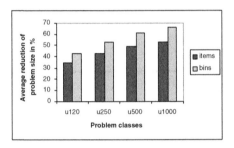

Fig. 1. Distribution of item sizes before and after preprocessing

Figure 1 shows the resulting reduction in problem size averaged over each problem class. Two observations can be made. First, problem sizes are significantly reduced through the preprocessing. The reduction ranges from 1/3 to 2/3 of the original problem size. Second, the percentage of reduced items is smaller than the percentage of the reduced bins [2]. This is due to the fact that the preprocessing eliminates mainly the large items. Moreover, the elimination of these large items makes the remaining problem easier to solve, due to both its smaller size and the fact that the remaining small items are potentially easier to place.

3   Computational Analysis

In this section we will turn to the description and analysis of the results obtained by our computational tests. Our AntPacking algorithm was coded in C and executed on a Pentium 3 with 750 MHz.

The problem instances used for the analysis of our algorithm are taken from Falkenauer (1996). The set consists of 80 problem instances randomly generated according to a uniform distribution with a bin capacity of 150 and item sizes

---

[2] Note, that the percentage of reduced bins can only be computed a posteriori, when a solution and the associated number of bins is known.

uniformly distributed between 20 and 100. The 80 problem instances are divided into 4 classes of 20 instances each. The classes differ with respect to problem size. More precisely, the smallest instances contain 120 items, and the other three classes feature instances with 250, 500 and 1000 items, respectively.

We tuned our algorithm in a preliminary experiment on some self made random instances and the following parameters have turned out to be appropriate: $\alpha = \beta = 1$, $q_0 = \rho = \tau_0 = 0.5$ and $z = 2$. The number of ants was fixed to 10 and the time limit for termination of the algorithm was set to 60 seconds. Preprocessing was used for all experiments. In these preliminary experiments we also verified the contribution of heuristic and pheromone information. It turned out that the lack of pheromone information has a much more detrimental effect to solution quality than the lack of the heuristic guidance. Further details about these tests are omitted here due to space limitations.

Two types of results will be presented in the following. First we will provide a detailed analysis of different design decisions for our algorithm considering pheromone decoding and update. Second, we will provide a benchmark comparison with two previous approaches, namely the Hybrid Grouping Genetic Algorithm (HGGA) of Falkenauer (1996) and the Hybrid ACO (HACO) of Levine and Ducatelle (2004).

Let us first turn to the design issues. To evaluate the merit of both our new pheromone decoding schemes and pheromone update strategies we have compared their performance with other, more standard schemes. Concerning the decoding, we compare our approach (which we will denote by $ol - bfg$ for object length - bin filling grade) with three other schemes. These are denoted by $o - bfg$, $o - ao$ and $ol - aol$. The $o - bfg$ approach relates an item (not its length) to the filling grade of a bin. The $o - ao$ scheme relates an item to all the items in the same bin. Finally, the $ol - aol$ scheme is the one used by Levine and Ducatelle (2004) and relates the length of an item with the lengths of all other items in the same bin. Thus, we have four different types of pheromone decoding.

Concerning the pheromone update we compare our local binary strategy (which we will refer to as lb) with four other strategies. The first one is the global strategy used by Levine and Ducatelle (2004), where only the best ant updates the memory (referred to as global (best ant)). A variation of this is the case where all ants may update the memory (global (all ants)). The third alternative is to compute the fitness for each bin individually and to have only the best ant update the memory (local (best ant)). Finally, the last alternative strategy is to use the local strategy, and to let all ants update the memory.

By considering all possible combinations, we get 20 different scenarios which we will compare based on two criteria. First, we measure the percentage of optimal solutions found. Second, we provide the average running times needed to return the best solutions (optimal or sub-optimal).

These results are summarized in Tables 1 and 2. Several observations concerning both, the pheromone update as well as the pheromone decoding strategies can be drawn from these tables. Let us start with the pheromone update strategies. First, we see that for both the local and the global pheromone update strategies it is always favorable to use only the best ant for updating. Using all

Table 1. Percentage of found optima

Pheromone update strategy	Pheromone decoding strategy			
	o-bfg	ol-bfg	o-ao	ol-aol
lb	98.75%	100%	98.75%	98.75%
local (all ants)	88.75%	92.5%	93.75%	81.25%
local (best ant)	98.75%	97.5%	98.75%	97.5%
global (all ants)	88.75%	91.25%	90%	80%
global (best ant)	98.75%	98.75%	98.75%	96.25%

ants performs worse for all three criteria. Second, there seems to be no systematic difference between the local and the global pheromone update strategies if only the best ant updates the memory. Third, the *lb* approach outperforms both the local and and the global approach as a pheromone update strategy with respect to both criteria.

Table 2. Average computation times to find the best solution (in seconds)

Pheromone update strategy	Pheromone decoding strategy			
	o-bfg	ol-bfg	o-ao	ol-aol
lb	1.23	1.4	1.68	2.07
local (all ants)	7.61	5.53	7.35	13.52
local (best ant)	1.94	2.65	2.60	3.37
global (all ants)	8.19	6.57	9.33	15.41
global (best ant)	1.96	2.12	3.36	4

Let us now turn to the pheromone decoding strategies. First, from Tables 1 and 2 we observe that overall, the $o - bfg$, the $ol - bfg$ and the $o - ao$ algorithms perform clearly better than the $ol - aol$ approach, which is clearly the worst for both criteria. Second, concerning the comparison between the other three algorithms we find that, if one disregards the (dominated) pheromone updating strategies, where all ants are allowed to reinforce their solutions, there is essentially no difference in the approaches particularly with respect to algorithm effectiveness. Concerning efficiency, the $o-bfg$ algorithm performs best, followed by the $ol-bfg$ version of our AntPacking algorithm. Third, considering only the best setting for the pheromone update strategy, namely the *lb* case, the $ol-bfg$ algorithm performs best as it finds optimal solutions for all problem instances. However, mind that for this *lb* strategy all pheromone decodings work well and the differences in the results become very small.

To summarize, we have the following findings.

– The *lb* pheromone update strategy clearly outperforms the other strategies. Reinforcing well filled bins regardless of the quality of their associated solutions seems to enhance the exploitation of building blocks that may lead to excellent solutions. On the other hand it provides the advantages of ex-

ploration for items that have not occurred in completely filled bins. Thus, it alleviates to some extent the problems associated with reinforcing poor parts of solutions just because they are part of good solutions. Moreover, the *lb* strategy is significantly faster than the other strategies in terms of the time needed to find the best solutions.
- The pheromone decoding strategies based on the bins' degree of filling lead to a faster convergence to the best solutions than the algorithms based on the portfolio of items in the bins. This is due to the more compressed structure of the pheromone information.
- Among the two strategies that relate items to the portfolio of items already in a bin, characterizing items by their lengths (and thus grouping equal sized items) is clearly dominated by looking at each individual item. This is true for both solution quality and computational effort.

Given these results, we will now use the best algorithm, namely the *AntPacking − lb − ol − bfg* for comparison with two state of the art metaheuristic techniques. These are the HGGA of Falkenauer (1996) and the HACO of Levine and Ducatelle (2004). Results in terms of number of optimal solutions found and average computational times needed to find the best solutions are provided for the four classes of problems in Table 3.

Table 3. Comparison of the AntPacking algorithm with the HGGA and the HACO

Problem class	HACO		HGGA		AntPacking	
	# of optima	seconds	# of optima	seconds	# of optima	seconds
$u120$	20	1	20	0.31	20	0.04
$u250$	18	52	20	0.75	20	0.58
$u500$	20	50	20	1.5	20	1.11
$u1000$	20	147	20	3.78	20	3.88

Note that we ran the HGGA on the same machine as our AntPacking algorithm, whereas the results for the HACO are taken from the paper of Levine and Ducatelle (2004). Thus, the running times for the HACO can not be compared directly to our computation times. Levine and Ducatelle used a Sun Blade 100 machine with an 502 MHZ UltraSparc-IIe processor. Further, the HGGA also uses some kind of preprocessing to reduce problem size. However, this is not reported in Falkenauer (1996) and we just had a program executable, so we could not run the HGGA without preprocessing. Thus, both AntPacking and HGGA were run with preprocessing, whereas HACO was run without.

Keeping that in mind we believe that our results clearly show the excellent performance of our AntPacking algorithm in all respects. Both our algorithm and the HGGA find optimal solutions for all instances, while the HACO is not able to find optimal solutions for 2 instances. Moreover, it seems to be slower by an order of magnitude.

Summarizing, the AntPacking algorithm is able to solve all instances in the test set to optimality, thus matching the HGGA in terms of solution quality. Furthermore, the two algorithms show no obvious difference in their computational requirements.

4   Conclusion

In this paper we have proposed a new ACO based approach for the one-dimensional bin packing problem. Through the use of a preprocessing procedure we were able to reduce the actual problem size faced by the ants by up to 67%, thus improving the efficiency of the algorithm significantly. Apart from that we have proposed both a new pheromone decoding scheme and a new pheromone update strategy, which were shown to improve algorithm performance further. Finally, we presented results obtained from a comprehensive computational study that highlight the excellent performance of the algorithm. In fact, our AntPacking approach clearly outperforms an existing ACO based algorithm and performs at least as good as the HGGA of Falkenauer (1996), which is considered to be the best algorithm for BPPs.

Acknowledgments. We wish to thank E. Falkenauer for providing his code of the HGGA for our computational comparison. We are also grateful for valuable comments from three anonymous referees.

References

1. Dorigo, M. and Di Caro, G.: "The Ant Colony Optimization metaheuristic". In: Corne, D. et al. (eds.): *New Ideas in Optimization*, McGraw Hill, UK, (1999), p. 11–32.
2. Dorigo, M. and Stuetzle, T.: *Ant Colony Optimization*, MIT Press, Cambridge, MA, (2004).
3. Falkenauer, E. and Delchambre, A.: "A genetic algorithm for bin packing and line balancing". *Proceedings of the IEEE 1992 International Conference on Robotics and Automation*, Nice, France, (1992).
4. Falkenauer, E.: "A hybrid grouping genetic algorithm for bin packing". *Journal of Heuristics* **2** (1996), p. 5–30.
5. Garey, M. R. and Johnson, D. S.: *Computers and Intractability: A guide to the theory of NP-Completeness*. Freeman, San Francisco, CA, (1979).
6. Gutjahr, W. J.: "ACO Algorithms with Guaranteed Convergence to the Optimal Solution". *Information Processing Letters* **82** (2002), p. 145–153.
7. Levine, J. and Ducatelle, F.: "Ant Colony Optimisation and Local Search for Bin Packing and Cutting Stock Problems". accepted for: Special Issue on Local Search, *Journal of the Operational Research Society*, (2004).
8. Martello, S. and Toth, P.: *Knapsack Problems, Algorithms and Computer Implementations*, Wiley & Sons, UK, (1999).

Scatter Search and Memetic Approaches to the Error Correcting Code Problem

Carlos Cotta

Dept. Lenguajes y Ciencias de la Computación, University of Málaga,
ETSI Informática, Campus de Teatinos, 29071 - Málaga, Spain
ccottap@lcc.uma.es

Abstract. We consider the problem of designing error correcting codes (ECC), a hard combinatorial optimization problem of relevance in the field of telecommunications. This problem is tackled here with two related techniques, scatter search and memetic algorithms. The instantiation of these techniques for ECC design will be discussed. Specifically, the design of the local improvement strategy and the combination method will be treated. The empirical evaluation will show that these techniques can dramatically outperform previous approaches to this problem. Among other aspects, the influence of the update method, or the use of path relinking is also analyzed on increasingly large problem instances.

1 Introduction

Telecommunications undoubtedly constitute one of the most prominent pillars upon which our present society rests. Its crucial importance is well captured in the numerous open research lines that are currently challenging the scientific community. Many of the tasks found in this area can be formulated as combinatorial optimization problems, e.g., assigning frequencies in radio link communications [1,2], designing telecommunication networks [3,4], or developing error correcting codes for transmitting messages [5,6] among others. In this work, we will focus precisely on the latter problem.

Roughly speaking, the development of an error correcting code (ECC) consists of designing a communication scheme for maximizing the reliability of information transmission through a noisy channel. This task admits several formulations. Here, we have considered the case of binary linear-block codes [7]. The design of such codes turns out to be very difficult. There exists no known algorithm for efficiently finding optimal solutions. The utilization of metaheuristic approaches is thus in order. In this sense, this problem has been treated in the literature with simulated annealing [8], genetic algorithms (GAs) [9,6], and hybrids thereof [5], with moderate success. The use of scatter search (SS) [10] and memetic algorithms (MAs) [11] will be considered here.

Unlike other metaheuristics, SS and MAs are concerned explicitly with incorporating problem knowledge in the representations and/or operators. Although this generally hinders the theoretical analysis of these techniques, it is undisputable from a pragmatic point of view that the resulting algorithms are usually

J. Gottlieb and G.R. Raidl (Eds.): EvoCOP 2004, LNCS 3004, pp. 51–61, 2004.
© Springer-Verlag Berlin Heidelberg 2004

highly effective in solving a plethora of problems. This work will discuss the deployment of these algorithms to the ECC problem. It will be shown that a drastic improvement with respect to sophisticated versions of other metaheuristics can be achieved.

2   The Error Correcting Code Problem

As discussed in the previous section, an ECC is aimed at maximizing the reliability of message transmissions through a noisy channel. This objective requires introducing some redundancy in the messages (i.e., using more bits than strictly necessary,) to increase the chances of recovering a message if some bits flip while traversing the channel. Of course, this redundancy has to be limited since the increased length of messages results in slower communication.

Let us assume that messages are expressed in sequences of characters from some alphabet Σ. In the context of the binary linear block codes, we would map each of these characters $c_i \in \Sigma$ to a sequence of n bits (or code word) w_i in order to transmit it. Upon reception of a n-bit sequence w, the character encoded could be recovered by looking for the closest –in a Hamming distance sense– valid code word. It is easy to see that if all code words are separated by at least d bits, any modification of at most $(d-1)/2$ bits in a valid code word can be easily reverted. Hence large d is sought.

It is possible to increase the value of d by considering larger values of n, but as stated above an upper bound of n has to be considered. Thus, we would be interested in maximizing d for a certain alphabet Σ, and a certain value of n. This way, an ECC problem instance is fully specified by a pair (n, M), where n is the number of bits in each code word, and M is the number of code words. Let $\mathbb{B} = \{0, 1\}$; the solution space for an ECC problem instance would comprise all sets $C = \{w_1, \cdots, w_M\}$, $w_i \in \mathbb{B}^n$, i.e., all combinations of M different n-bit sequences. The size of the search space is thus $\binom{2^n}{M}$. Although no known algorithm is available for producing an optimal ECC (i.e., a set of M n-bit code words with maximal d) in general, the problem has been theoretically studied, and bounds on the attainable values of d for different combinations of n and M have been derived [12].

It is interesting to notice the relation between the ECC problem as defined above, and another –apparently unrelated– problem in the realm of physics: finding the lowest energy configuration on M particles in a n-dimensional space. By assimilating particles to code words, the ECC problem can be viewed as distributing M code words in the corners of a binary n-dimensional space. This connection was used by Dontas and de Jong [6] to define a fitness function (to be maximized) for a genetic algorithm optimizing this problem, i.e.,

$$Fitness(C) = \frac{1}{\sum_{i=1}^{M} \sum_{j=1, i \neq j}^{M} \frac{1}{d_{ij}^2}} , \tag{1}$$

where d_{ij} is the Hamming distance between code words w_i and w_j. This function is more adequate as a guiding function than a naïve function computing the minimum distance between different code words in a solution. Although the latter would capture the absolute quality of a solution, it would induce large plateaus in the fitness landscape. This would not be the case for the former function, which is capable of grasping the effects of small changes in a solution.

3 Scatter Search and Memetic Algorithms for the ECC Problem

SS is a metaheuristic based on populational search whose origin can be traced back to the 1970s in the context of combining decision rules and problem constraints. Unlike other populational approaches such as genetic algorithms, SS relies more on deterministic strategies rather than on randomization. This distinctive methodological difference notwithstanding, SS shares some crucial elements with MAs such as the use of combination procedures and local-improvement strategies. More precisely, the following components are present in the algorithmic template of SS:

- A *diversification generation method* for generating a collection of raw solutions, possibly using some initial solution as "seed".
- An *improvement method* for enhancing the quality of raw solutions.
- A *reference set update method* for building the reference set (i.e., the population) from the initial set of solutions generated, and for maintaining it by incorporating some solutions produced in subsequent steps.
- A *subset generation method* for selecting solutions from the reference set, and arranging them in small groups (pairs, triplets, or larger groups) for undergoing combination.
- A *solution combination method* for creating new raw solutions by combining the information contained in a certain group of solutions.
- A *restart reference set method* for refreshing the reference set once it has been found to be stagnated. This can be done by using the diversification generation method plus the improvement method mentioned above, but other strategies might be considered as well.

The design of a particular SS algorithm is completed once the items above are detailed. Next subsections will be devoted to this purpose. Notice at this point that the very same components cited before can be found in a MA (see e.g. [13];) the main difference between SS and MAs lies in the use of randomization in the latter (in particular, this implies the presence of a mutation operator, absent in SS.) This issue will be discussed in the next subsections as well.

3.1 Diversification Generation Method

The diversification generation method serves two purposes in the SS algorithm considered: it is used for generating the initial population from which the reference set will be initially extracted, and it is utilized for refreshing the reference set whenever a restart is needed.

The generation of new solutions is performed by using a randomized procedure that tries to generate diverse solutions, and whose code words are expected to be distant. To do so, a procedure loosely inspired in naïve Bayesian methods is utilized. More precisely, a count of the number of 1s appearing in each position of a code word is maintained, and used to bias the generation of bits for the next code word. The exact pseudocode of the algorithm is as follows:

1. $sol \leftarrow \emptyset$
2. **for** $i \in [1:n]$ **do** $c_i \leftarrow 0$
3. **for** $j \in [1:M]$ **do**
   a) **repeat**
      - **for** $i \in [1:n]$ **do** $w[i] \leftarrow \left(\text{URand01}() > \frac{c_i}{j}\right)$
      until $w \notin sol$
   b) **for** $i \in [1:n]$ **do** $c_i \leftarrow c_i + w[i]$
   c) $sol \leftarrow sol \cup \{w\}$

By using this procedure, the higher the frequency of 1s (resp. 0s) in a certain bit of the code words generated so far in a solution, the higher the chances that the next code word will have a 0 (resp. 1) in this bit.

3.2   Improvement Method

The improvement method is responsible for enhancing raw solutions produced by the diversification generation method, or by the solution combination method. In general, this is achieved by applying small changes to a solution, keeping them if they produce a quality increase, or discarding them otherwise.

These small changes amount in this case to the modification of single bits in a code word. This procedure benefits from the separability of the fitness function: the new solution is accepted if $\sum_{j=1,i\neq j}^{n} d'^{-2}_{ij} < \sum_{j=1,i\neq j}^{n} d^{-2}_{ij}$, where d_{ij} are the distances of the original code word w_i being modified to the remaining code words, and d'_{ij} are the distances of the modified code word w'_i. The whole process would be as follows:

1. **repeat**
   a) change←**false**
   b) **for** $j \in [1:M]$ **do**
      i. Find the two closest code words, such that at least one of them has not been modified yet. Let w be the unmodified codeword (or the code word with the lowest index if both are unmodified.)
      ii. **for** $i \in [1:n]$ **do**
         A. $w[i] \leftarrow (1 - w[i])$
         B. **if** the change is acceptable **then** retain it, and set change←**true**
            else undo it
   until ¬change

The underlying idea of this procedure is trying always to separate the closest code words, aiming at maximizing the minimal distance d. Notice that changes resulting in lower values of the fitness function, are reverted as indicated in step 1(b)iiB.

3.3   Subset Generation Method

This method generates the groups of solutions that will undergo combination. A binary combination method has been considered in this work, and hence this subset generation method forms couples of solutions. Following the SS philosophy, this is done exhaustively, producing all possible pairs. It must be noted that since the combination method utilized is deterministic, it does not make sense to combine again pairs of solutions that were already coupled before. The algorithm keeps track of this fact to avoid repeating computations.

In the case of the MA, the selection mechanism plays the role of this subset generation method. As it is typically done, a fitness-based randomized selection method has been chosen. More precisely, binary tournament is used to select the solutions that will enter the reproductive stage.

3.4   Solution Combination Method

This method is fed with the subsets generated by the previous method, and produces new trial solutions by combining the information contained in each of these subsets. Two different alternatives have been considered for this method, a greedy combination procedure, and path relinking.

The greedy combination procedure (GR) incrementally constructs a new solution by greedily selecting code words from the parents, i.e., at each step the code word w that maximizes $D(w) = \sum_{j=1}^{k} d_{(w)j}^{-2}$ is taken, where k is the current number of code words in the solution. More precisely, let sol_1 and sol_2 be the solutions being combined; the pseudocode of the process is:

1. $newsol \leftarrow sol_1 \cap sol_2$
2. $candidates \leftarrow (sol_1 \cup sol_2) \setminus (sol_1 \cap sol_2)$
3. **while** $|newsol| < M$ **do**
   a) Pick $w \in candidates$ such that w minimizes $D(w)$
   b) $newsol \leftarrow newsol \cup \{w\}$
   c) $candidates \leftarrow candidates \setminus \{w\}$

As it can be seen, this combination procedure is designed to respect common code words, present in both parents. This helps focusing the search, by promoting exploitation. In the case of the MA, a randomized version of this combination method has been devised. To do so, step 3a is modified so as to pick the ith best candidate with probability proportional to 2^{-i}.

The alternative to GR is path relinking (PR) [14]. This method works by generating a path from an initiating solution to a destination solution. At each step of the path, a new solution is generated by substituting a code word absent from the destination solution by a code word present in the latter. The code word to be substituted is here selected in a greedy fashion. The best solution in the path (excluding the endpoints, already present in the reference set) is returned as the output of the combination procedure:

1. $current \leftarrow sol_1$; $bestfit \leftarrow 0$; $bestsol \leftarrow sol_1$
2. $candidates \leftarrow sol_2 \setminus sol_1$
3. **while** $|candidates| > 1$ **do**
   a) Pick $w \in sol_1 \setminus sol_2$ such that w maximizes $D(w) - D(w')$, where w' is the code word with the lowest index in $candidates$
   b) $current \leftarrow newsol \cup \{w'\} \setminus \{w\}$
   c) $candidates \leftarrow candidates \setminus \{w'\}$
   d) **if** $Fitness(current) > bestfit$ **then**
      i. $bestsol \leftarrow current$
      ii. $bestfit \leftarrow Fitness(current)$

The above procedure can be augmented by applying the improvement method to *bestsol* whenever it is updated. This strategy is termed PR-LS.

3.5   Reference Set Update Method

The reference set update method must produce the reference set for the next step by using the current reference set and the newly produced offspring (or by using the initial population generated by diversification at the beginning of the run or after a restart.) Several strategies are possible here. Quality is an obvious criterion to determine whether a solution can gain membership to the reference set: if a new solution is better than the worst existing solution, the latter is replaced by the former. Notice the similarity with the plus replacement strategy commonly used in other evolutionary algorithms. This plus strategy has been precisely considered in the MAs used in this work.

A variant of this update method has been also considered: rather than generating all descendants and then deciding which of them will be included in the reference set, descendants can be generated one-at-a-time, and inserted in the reference set if they qualify for it. This is called a dynamic updating as opposed to the static updating described before. As it can be seen, the dynamic updating resembles a steady-state replacement strategy, while the static updating would be similar to a generational model. We will thus assimilate steady-state MAs –i.e., $(\mu_{MA} + 1)$– to dynamic updating, and elitist generational MAs –i.e., $(\mu_{MA} + \mu_{MA})$– to static updating, where μ_{MA} is the MA population size.

3.6   Restart Reference Set Method

The restart method must refresh the reference set by introducing new solutions whenever all pairs of solutions have been coupled without yielding improved solutions. This is done in our SS algorithm as follows: let μ_{SS} be the size of the reference set; the best solution in the reference set is preserved, $\lambda_{SS} = \mu_{SS}(\mu_{SS} - 1)/2$ solutions are generated using the diversification generation method and the improvement method, and the best $\mu_{SS} - 1$ out of these λ_{SS} solutions are picked and inserted in the reference set.

Restarting is also possible in MAs, although given the randomized nature of the operators in this case, some ad-hoc criterion must be used to determine the existence of a diversity crisis (e.g., some statistical analysis of the population.) A simpler alternative has been utilized in this work: rather than using a full restarting method, a larger population and a mutation operator to permanently inject diversity have been considered. Concretely, the MA population comprises $\mu_{MA} = \lambda_{SS}$ solutions. As to the mutation operator, standard bit-flipping is used.

4   Computational Results

The experiments have been realized using a reference set of $\mu_{SS} = 10$ solutions. Hence, the MA has a population of $\mu_{MA} = 45$ solutions. Other parameters of the MA are the probability of recombination $p_X = 0.9$, and the probability of mutation $p_m = 1/\ell$, where $\ell = n \cdot M$ is the total number of bits in solutions.

The first test has been done on a 12-bit/24-word ECC problem instance. This is the same problem instance considered in other works in the literature, and thus allows comparing the relative performance of SS and MA. Table 1 shows the results. An important observation with respect to these results is that the number of evaluations reported includes the partial calculations performed during local improvement or combination. This has been done by considering the computation of the distance between two code words as the basic unit; whenever such a calculation is done, an internal counter is incremented. By dividing the value of this counter by $M(M-1)/2$ we obtain the equivalent number of additional full evaluations performed. This way, fair comparisons are possible.

Table 1. Results (averaged for 50 runs) of the different variants of SS (greedy recombination –GR–, path relinking –PR–, and path relinking with local search –PR-LS–) and MA on a 12-bit/24-word ECC problem instance.

			number of evaluations			
		% opt.	min	mean ± std.	max	median
SS-static	GR	100%	3889	11313.62 ± 3388.94	19930	11791.5
	PR	100%	3889	10850.38 ± 2807.92	18535	11191
	PR-LS	100%	3889	11731.54 ± 4110.32	21616	11722
SS-dynamic	GR	100%	3889	8092.02 ± 1420.95	11928	7828
	PR	100%	3889	11460.18 ± 6760.63	42042	9235
	PR-LS	100%	3889	9586.58 ± 4476.21	38358	8646
MA	gen. elitist	100%	3889	15636.08 ± 6779.71	35129	15157.5
	steady-state	100%	3889	13416.16 ± 4395.62	23847	13275.5

As Table 1 indicates, all versions of SS and MA are capable of solving to optimality ($d = 6$) the problem in 100% of the runs, and in a small number of evaluations. To put these results in perspective, consider other results reported in the literature. In [5], a massively parallel genetic simulated annealing algorithm (parGSA) requires 16,384 processing elements just to achieve a performance similar to that of the MA (for 256 processing elements, the number of evalua-tions required by parGSA tops 30,000.) In [9], different sequential and parallel GAs are tested; despite using a knowledge-augmented representation (individ-uals only comprise 12 code words; the remaining 12 are obtained by inverting these,) steady-state and generational GAs only achieve 40% and 10% success re-spectively. A cellular GA achieves 100% success but requires 52,757 evaluations on average, far more than MAs or SS. Distributed versions of the steady-state and cellular GAs are capable of 100% success with 8 subpopulations, requiring 36,803 and 89,598 evaluations on average respectively. Again this is substantially higher than the results of SS and MAs.

Table 2. Statistical significance of the difference in number of evaluations to find the optimal solution for the 12-bit/24-word ECC problem instance. For each pair of algorithms, there are four symbols corresponding to the comparison of static vs static, dynamic vs dynamic, static vs dynamic, and dynamic vs static. ('+' = significant; '−' = not significant). The last column compares the static and dynamic versions of the same algorithm.

	SS/GR	SS/PR	SS/PR-LS	MA	stat. vs dyn.
SS/GR	•	−,+,−,+	−,+,+,+	+,+,+,+	+
SS/PR	−,+,+,−	•	−,−,+,−	+,+,+,+	−
SS/PR-LS	−,+,+,+	−,−,−,+	•	+,+,−,+	+
MA	+,+,+,+	+,+,+,+	+,+,+,−	•	−

Table 2 presents the statistical significance of the differences in number of evaluations. The Wilcoxon ranksum test (also known as Mann-Whitney U test) [15] has been used for this purpose. This test does not assume normality of the samples (as for example t-test does.) Such an assumption would be unrealistic for this data. As Table 1 shows, the dynamic versions provides better results than static ones in general. This difference is significant for SS/GR and SS/PR-LS. The different versions of SS also appear to be better than MAs, and this difference is in general significant. The best results are provided by the dynamic SS/GR, and this superiority is always significant.

Further experiments have been conducted in order to test the scalability of the algorithms. To be precise, 16-bit/32-word and 20-bit/40-word ECC problem instances have been considered. In this case, the algorithms have been allowed a maximum number of evaluations of $1.25 \cdot 10^6$ and $2.5 \cdot 10^6$ respectively. The results are shown in Tables 3 and 5.

In the 16-bit/32-word ECC problem instance, the results are very satisfac-tory: 100% success ($d = 8$) is achieved almost always. The SS/PR algorithm and the static MA are all above 95% success. Precisely the SS/PR algorithm appears to be slightly worse than the remaining SS algorithms. The superiority of the latter ones over SS/PR is always significant, as shown in Table 4. On the other hand, there is no significant difference among SS/GR and SS/PR-LS, both

Table 3. Results (averaged for 50 runs) of the different variants of SS and MA on a 16-bit/32-word ECC problem instance.

			number of evaluations			
		% opt.	min	mean ± std.	max	median
SS-static	GR	100%	28131	62826.32 ± 45930.51	338659	51628.5
	PR	98%	34155	149521.48 ± 145023.36	780843	100716
	PR-LS	100%	31212	79429.12 ± 153135.40	1145040	54990
SS-dynamic	GR	100%	12625	60590.86 ± 48997.89	243698	48444
	PR	96%	17197	216989.29 ± 259101.23	1173888	105140.5
	PR-LS	100%	13958	97871.24 ± 126648.54	768707	42623
MA	gen. elitist	98%	53571	89982.12 ± 17836.87	145423	91563
	steady-state	100%	35346	59640.88 ± 15557.11	101743	60499

Table 4. Statistical significance of the difference in number of evaluations to find the optimal solution for the 16-bit/32-word ECC problem instance.

	SS/GR	SS/PR	SS/PR-LS	MA	stat. vs dyn.
SS/GR	•	+,+,+,+	−,−,−,−	+,−,−,+	−
SS/PR	+,+,+,+	•	+,+,+,+	−,+,+,−	−
SS/PR-LS	−,−,−,−	+,+,+,+	•	+,−,−,+	−
MA	+,−,+,−	−,+,−,+	+,−,+,−	•	+

static and dynamic. As to the MA, the dynamic version provides significantly better results than the static MA, and similar to those of SS/GR or SS/PR-LS.

The algorithms start to exhibit the effects of the increased dimensionality on the 20-bit/40-word ECC problem instance. The success ($d = 10$) ratio clearly drops in this case. As for the previous instance, the SS/PR provides the worst results with a mere 4% success. The SS/GR algorithm yields the best results, with a success ratio about two or three times higher than the remaining algorithms. The MA provides a more or less similar success ratio than that of the SS/PR-LS. Since the success ratio of the different algorithms is very disparate, a full statistical comparison of the average number of evaluations would not make much sense here. At any rate, notice that, despite the MA has not a high success ratio, it provides the best results in number of evaluations, and this result is statistically significant against SS/GR and SS/PR-LS (except for the dynamic version of the latter.) This indicates that it can find relatively fast the optimal solution, but in most runs it stagnates in some locally optimal region of the search space. This suggests the need for using here a restarting method, since mutation alone cannot provide enough diversity in long runs of the MA on this problem instance.

Table 5. Results (averaged for 50 runs) of the different variants of SS and MA on a 20-bit/40-word ECC problem instance.

			number of evaluations				
		% opt.	min	mean ± std.		max	median
SS-static	GR	58%	129248	1018434.31 ±	675830.06	2378854	1025514
	PR	4%	257118	517757.00 ±	260639.00	778396	517757
	PR-LS	26%	143377	1092756.46 ±	588769.83	1957000	1080258
SS-dynamic	GR	54%	176121	1029429.78 ±	690248.82	2416476	973267
	PR	4%	129268	1143246.00 ±	1013978.00	2157224	1143246
	PR-LS	16%	133300	860126.00 ±	702699.19	2351918	770137.5
MA	gen. elitist	16%	181503	248624.25 ±	33230.08	284643	266163
	steady-state	18%	114339	134901.56 ±	22204.43	190900	130779

5   Conclusions

The results presented in the previous section clearly indicate that SS and MAs are cutting-edge techniques for solving the ECC problem, capable of outperforming sophisticated versions of other metaheuristics on this domain. In general, SS appears to be somewhat better than MAs, specifically when using GR. This latter method has shown to provide better results than PR or PR-LS. As to the update method, the dynamic version usually provides better results both in MAs and SS, although the difference is not always significant.

Future work will focus on improving some aspects of the algorithms. In the case of SS, the update method can consider additional criteria besides quality, such as diversity for instance. This implies structuring the reference set in several tiers (see [10] for details,) and can be useful for tackling larger instances. In the case of the MA, the addition of a full restart method is likely to produce remarkable improvements in those larger instances as well.

Acknowledgements. This work is partially supported by Spanish MCyT, and FEDER under contract TIC2002-04498-C05-02.

References

1. Dorne, R., Hao, J.: An evolutionary approach for frequency assignment in cellular radio networks. In: 1995 IEEE International Conference on Evolutionary Computation, Perth, Australia, IEEE Press (1995) 539–544
2. Kapsalis, A., Rayward-Smith, V., Smith, G.: Using genetic algorithms to solve the radio link frequency assigment problem. In Pearson, D., Steele, N., Albretch, R., eds.: Artificial Neural Nets and Genetic Algorithms, Wien New York, Springer-Verlag (1995) 37–40
3. Chu, C., Premkumar, G., Chou, H.: Digital data networks design using genetic algorithms. European Journal of Operational Research **127** (2000) 140–158

4. Vijayanand, C., Kumar, M.S., Venugopal, K.R., Kumar, P.S.: Converter placement in all-optical networks using genetic algorithms. Computer Communications **23** (2000) 1223–1234

5. Chen, H., Flann, N., Watson, D.: Parallel genetic simulated annealing: A massively parallel SIMD algorithm. IEEE Transactions on Parallel and Distributed Systems **9** (1998) 126–136

6. Dontas, K., Jong, K.D.: Discovery of maximal distance codes using genetic algorithms. In: Proceedings of the Second International IEEE Conference on Tools for Artificial Intelligence, Herndon, VA, IEEE Press (1990) 905–811

7. Lin, S., Jr., D.C.: Error Control Coding : Fundamentals and Applications. Prentice Hall, Englewood Cliffs, NJ (1983)

8. Gamal, A., Hemachandra, L., Shaperling, I., Wei, V.: Using simulated annealing to design good codes. IEEE Transactions on Information Theory **33** (1987) 116–123

9. Alba, E., Cotta, C., Chicano, F., Nebro, A.: Parallel evolutionary algorithms in telecommunications: Two case studies. In: Proceedings of the CACIC'02, Buenos Aires, Argentina (2002)

10. Laguna, M., Martí, R.: Scatter Search. Methodology and Implementations in C. Kluwer Academic Publishers, Boston MA (2003)

11. Moscato, P.: Memetic algorithms: A short introduction. In Corne, D., Dorigo, M., Glover, F., eds.: New Ideas in Optimization. McGraw-Hill, London UK (1999) 219–234

12. Agrell, E., Vardy, A., Zeger, K.: A table of upper bounds for binary codes. IEEE Transactions on Information Theory **47** (2001) 3004–3006

13. Moscato, P., Cotta, C.: A gentle introduction to memetic algorithms. In Glover, F., Kochenberger, G., eds.: Handbook of Metaheuristics. Kluwer Academic Publishers, Boston MA (2003) 105–144

14. Glover, F., Laguna, M., Martí, R.: Fundamentals of scatter search and path relinking. Control and Cybernetics **39** (2000) 653–684

15. Lehmann, E.: Nonparametric Statistical Methods Based on Ranks. McGraw-Hill, New York NY (1975)

A Hybrid Evolutionary Algorithm for Solving the Register Allocation Problem

Betul Demiroz[1], Haluk Topcuoglu[1], and Mahmut Kandemir[2]

[1] Computer Engineering Department, Marmara University, 81040, Istanbul, Turkey
{bdemiroz,haluk}@eng.marmara.edu.tr
[2] Dept. of Computer Science and Engineering, Pennsylvania State University
University Park, PA 16802
kandemir@cse.psu.edu

Abstract. One of the strong impacts on runtime performance of a given code is the register allocation phase of compilation. It is crucial to provide aggressive and sophisticated register allocators for the embedded devices, where the excessive compilation time is tolerated because of its off-line nature. In this paper, we present a hybrid evolutionary algorithm for register allocation problem that combines genetic algorithms with a local search technique. Our algorithm exploits a novel, highly-specialized crossover operator that considers domain-specific information. Computational experiments performed on various synthetic benchmarks prove that our method significantly outperform the state-of-the-art algorithms with respect to all given comparison metrics on solution quality.

1 Introduction

With the proliferation of small factor embedded devices such as two-way pagers, cellular phones, retailer terminals, and PDAs, optimizing embedded software is becoming increasingly important. While hand-optimized embedded code can result in high performance and low power/memory footprint, it is a tedious, complex task and error-prone. Consequently, current trends are toward automatically optimizing code using compiler support. In particular, an optimizing compiler can analyze data access patterns exhibited by the application, and restructure code and/or data to improve its performance, memory footprint, and/or power consumption. Register allocation is one of the crucial parts of a compiler as it determines how program variables should be assigned to architectural registers. State-of-the-art register allocators usually employ heuristic solutions since the general register allocation problem is known to be NP-hard [10]. While this maybe acceptable for a general-purpose system where each day many programs of different types are compiled and executed, one can potentially do much better in embedded systems. There are two major reasons for this argument. First, the number of application programs that will execute on a typical embedded system is typically small (usually dictated by the application domain of interest). For example, a microcontroller in an automobile typically runs a single application throughout its lifetime. Similarly, a PDA runs a small set of applications.

J. Gottlieb and G.R. Raidl (Eds.): EvoCOP 2004, LNCS 3004, pp. 62–71, 2004.

Therefore, it may be wise to tailor the compilation process to the needs of the application domain. Second, one can spend a lot of machine cycles in compiling an application for embedded systems as the code quality is very critical. As a result, it is important to search for more aggressive/sophisticated register allocators that can take lots of time to execute but eventually generate high-quality code that outperforms state-of-the-art heuristic-based solutions. The excessive compilation time is usually not a major problem as compilation is an off-line process.

In this paper, we study how Genetic Algorithms (GA) can be used for deriving high-quality register allocators. We contribute a highly specialized crossover operator called Conflict-Free Partition Crossover (CFPX) that incorporates the specific information on register allocation problem. The CFPX operator combined with an efficient local search method leads to a very powerful method, which resides in the class of hybrid evolutionary algorithms. Our experimental evaluation through a set of carefully engineered synthetic benchmarks clearly indicate that the proposed solution outperforms a widely-used register allocation heuristic [1,2] over a diverse set of execution parameters.

The remainder of this paper is organized as follows. The next section discusses the main issues regarding the register allocation problem. Section 3 presents the details of our hybrid algorithm. We present the computational experiments in Section 4. The conclusions are given in Section 5.

2   Register Allocation Problem

A register allocator of a compiler maps symbolic registers (variables) into architectural registers; and values stay in registers as long as they are live. A live range is the union of the regions in the program in which the symbolic registers are live. The graph coloring paradigm has been considered for performing the register allocation and Chaitin [3] designed and implemented the first graph-coloring allocator. To model register allocation as a graph coloring problem, the compiler first constructs an interference graph G, where the nodes represent the live ranges of the symbolic registers and edges represent the interferences. If any two live ranges l_i and l_j are simultaneously live at some point, they cannot be stored in the same architectural register. Therefore, there will be an edge between nodes i and j in graph G. Two terms, live range of a variable and its corresponding node in the graph, are interchangeably used in this paper. After the interference graph is constructed, the compiler looks for an assignment of k colors to the nodes of G with distinct colors for adjacent nodes, where k is the number of available architectural registers. If k-coloring is not possible, one or more variables are kept in memory rather than in architectural registers, which is called spilling. The spill cost of a live range is the estimated run-time cost of the corresponding symbolic register for loading from and storing in memory. A register allocation method targets to minimize the total spill cost value; and in the best case total spill cost is equal to zero.

Even the simple formulations of graph coloring problems are NP-complete [10], a large set of heuristic approaches are presented in the literature. The recent efforts [7,9,11] on graph coloring problem is to propose hybrid evolutionary algorithms that include problem-specific crossovers with local search methods.

The hybrid evolutionary algorithms presented for graph coloring problem do not fully consider the characteristics of the register allocation problem. Specifically, the register allocation problem target to allocate the registers for the variables that have higher spill costs so that it minimizes the overall spill cost. However, the graph coloring problem do not differentiate between two vertices if they have the same number of outgoing edges.

2.1   Related Work

The first implementation of a register allocator based on graph coloring was the Chaitin's heuristic [3]. A later strategy based on a coloring paradigm is the priority-based method described by Chow and Hennessy [4]. It was given in [2] that all subsequent work on graph coloring allocators has derived from one of these two strategies.

On the other hand, the Chaitin's heuristic is not guaranteed to find a k-coloring if it exists. As an example, the diamond graph with four nodes was used in the literature to present the limitations of the Chaitin's heuristic, where Chaitin's heuristic requires spilling for this graph. The optimistic coloring heuristic (OCH) [1,2], a Chaitin-style graph coloring register allocator, generates 2-coloring of the diamond graph.

At each step of OCH, a node from the interference graph that has degree less than k is selected randomly and it is put on the stack. Then, the selected node and all edges incident to it in graph G are removed. This process is repeated until either graph G becomes empty or all remaining nodes have degree equal to or greater than k. In the latter case, a node is selected as a spill candidate using a predefined technique. The most common technique given in the literature selects the node that minimizes the following equation

$$h(i) = \frac{SCost(i)}{C\_Degree(i)}, \tag{1}$$

where $SCost(i)$ and $C\_Degree(i)$ are the spill cost and the current degree of the node i in G. The selected node and its edges are removed from the graph and the node is pushed into the stack, instead of marking it for spilling. This process is repeated until graph G becomes empty.

Then, the nodes are popped from the stack successively; and each popped node is assigned to the color (i.e., the register) with the lowest index that is not used by any of its neighbor. If no color is available for a node, it is left uncolored. The uncolored nodes are spilled and the summation of spill costs of the uncolored nodes is the total spill cost of the solution.

3   A Hybrid Evolutionary Algorithm for Register Allocation

A hybrid evolutionary algorithm (HEA) integrates Genetic Algorithms (GA) with local search techniques [5,6]. The general procedure of our hybrid algorithm is as follows. The algorithm sets the initial population of solutions, which is followed by a number of iterations. At each iteration, two strings are chosen from the population at random. Then, the problem-specific crossover operator is applied to generate a single offspring, which is followed by the local search operator to improve the offspring before inserting it into the population. This process repeats until a predefined number of iterations is reached. In our GA-based approach, we consider the partition method [9] for string representation. Each solution S_i partitions the variables into register classes $S_i = \{R_1, R_2, ...R_k\}$ where each class R_i includes the live ranges of variables that are mapped to the register r_i; and k is the total number of registers in the system.

3.1   Initial Population Generation

A predefined number of strings are generated as part of initial population generation step. For this step, we modified the DSatur Algorithm [8], which is a graph coloring heuristic. The algorithm is so called because of using saturation degree of each node, which is defined by the number of different colors to which the node is adjacent to. The DSatur algorithm arranges vertices by decreasing order of saturation degrees. In our method, we define a new term called spill degree for each node i, which is set with

$$S\_Degree(i) = \frac{S\_Cost(i)}{Degree(i)}, \tag{2}$$

where $S\_Cost(i)$ is the spill cost and $Degree(i)$ is the number of edges incident to the node for the variable i. This equation is similar to the Equation 1 given in Section 2. The optimistic graph coloring uses minimization to choose a variable for spilling, whereas we use maximization to choose a variable for assigning to a register class.

In our method, nodes are sorted in decreasing order of spill degrees. For each iteration, the uncolored node with the maximal spill degree is mapped to the register class with lowest possible index value where the selected node should be conflict-free with the other nodes in the same class. (If there is no edge between any two nodes i and j in the interference graph, then nodes i and j are conflict-free.) In case of tie-breaking for spill degree values, the node with the maximal degree is selected. If it is not possible to locate a node, the algorithm selects a register class at random. This process is repeated until all variables are mapped to one of the register classes.

3.2   Crossover Operator

We introduce a novel crossover operator in order to account for problem-specific information. The crossover operator used in our implementation is called Conflict-Free Partition Crossover (CFPX), as it targets to provide conflict-free register partitions. The crossover operator constructs an offspring successively, by generating a single partition (or class) of the offspring at each step.

Assume that two parents, $S_1 = \{R_1^1, ..., R_k^1\}$ and $S_2 = \{R_1^2, ..., R_k^2\}$, are selected for the crossover randomly. The partition from these parents that has the maximum number of conflict-free nodes is selected. In case of tie-breaking for selecting a partition within a parent, a register class is selected randomly. If tie-breaking is between two parents, the register class is selected from the parent which is not considered in the previous step.

Assume that R_i, the i'th register class from one of the parents, is selected. The conflict free subset of R_i represented by R_i^{CF} becomes the base of the first register class for the offspring. Then, this base set is extended with other nodes by preserving its conflict-free property. The end of this process determines the first register class of the offspring and the same process is repeated for the other register classes. If a variable n is not assigned to any class at the end of this process, n is assigned to the class that has the minimum number of conflicting nodes with the variable n.

The process of determining the conflict-free set of each register class R_i is given in Figure 1. Initially, the conflict free set includes all variables in R_i. The total number of conflicts of each variable m in class R_i, $CF(m)$, is determined. The summation of $CF(m)$ values gives the total number of conflicts in the register class R_i. Then, the heuristic removes variables from the partition one by one in the decreasing order of $CF(m)$ values until the total number of conflicts in R_i equals to the half of its initial value.

Initialize conflict-free set R_i^{CF} of register class R_i with the set R_i.
Compute $CF(m)$ (total number of conflicts) for each variable m in R_i.
$CF^{all} = \sum_{m \in R_i} CF(m)$.
Initialize j to 0.
while ($j < \frac{C_i^{all}}{2}$)
    Select the variable n where $CF(n) = \max_{m \in R_i^{CF}} \{ CF(m) \}$.
    In case of tie-breaking, select the variable n where
        $S\_Cost(n) = \min_{m \in R_i^{CF}} \{ S\_Cost(m) \}$.
    Remove variable n from R_i^{CF}.
    $j = j + CF(n)$.
endwhile

Fig. 1. The Heuristic For Determining Conflict-Free Set of a Register Class

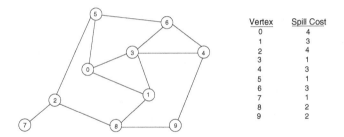

Vertex	Spill Cost
0	4
1	3
2	4
3	1
4	3
5	1
6	3
7	1
8	2
9	2

Fig. 2. An Example Interference Graph with 10 Vertices with the Spill Costs

Figure 3 is an example of the crossover operator applied on two parents derived from the interference graph given in Figure 2. When the first parent P1 is examined, the conflict-free subsets of the four partitions become $\{3\}$, $\{0, 6\}$, $\{7, 8\}$ and $\{9, 1, 5\}$, respectively. The maximum number of conflict-free nodes in classes of P1 becomes 3, which is equal to that of parent P2 due to its third register class. The total spill cost of P1 is equal to 7 due to the nodes 6 (or 4) and 2. For P2, the total spill cots equals to 10, which is due to the nodes 3, 1, 2 and 9. Then, the set ($\{9, 1, 5\}$) becomes base of the first register class of the result. This set is extended with node 7, since there is no edge between 7 and any node from the set. The complete offspring is generated when this process is repeated for the other register classes.

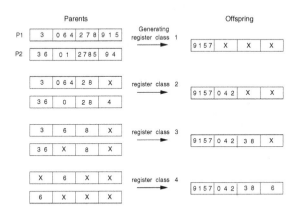

Fig. 3. An Example of Applying the Crossover Operator

A similar crossover operator is the Greedy Partition Crossover (GPX) given in [9]. It considers the parents successively but it does not consider the conflicts while selecting the partitions. Additionally, it does not have a second phase of extending the partition with the variables that belong to other partitions by preserving the conflict-free property.

3.3   Local Search Phase

After a solution is generated using the crossover operator, the local search phase targets to improve it by checking a set of its neighborhood solutions. A solution S_2 is a neighbor of S_1 if the two solutions differ in a single location of a variable i in the register classes and i is in conflict with one or more variables in its register class of S_1. In this phase, we compute the spill factor of each variable i using the following equation:

$$S\_Factor(i) = S\_Cost(i) \times CF(i) \qquad (3)$$

Then, the variable i which has the maximum spill factor is selected and it is moved to another register class where its total number of conflicts decreases. The search for target partition is started with the register class which has the minimum number of variables and it is repeated in increasing order of class sizes until a decrease in $CF(i)$ is achieved. This search process stops in a class which causes a decrease in number of conflicts of variable i. If it is not possible to provide a decrease, the search process is repeated for the variable with the next maximal spill factor value.

3.4   Fitness Function

For each register class R_i of a given solution, the total number of conflicts of each variable j in R_i is determined. Then, the variables in R_i are put in a spill set and removed from R_i in decreasing order of their number of conflicts until the class R_i becomes conflict-free. In case of tie-breaking, the one with the minimum spill cost value is put into the spill set. The summation of spilled costs in this set gives the partial fitness value (i.e. for R_i). This process is repeated to all register classes in the solution; and the overall summation of fitness values of classes gives the fitness value of the solution.

4   Experimental Study

We implemented a graph generator for developing random interference graphs, which requires three input parameters: the number of nodes (n), edge density (α) and the mean spill cost (γ) for the variables. The edge density is the probability that an edge is present between any two distinct nodes, where we assume a uniform probability distribution. After a graph is constructed, the spill cost of each node is set randomly with a uniform distribution in the range $[1..2 \times \gamma]$. Each generated graph is tested with various number of registers. A register density (β) term is introduced, where the total number of registers is equal to $r = \beta \times n$. A set of experiments are conducted for identifying the good parameter settings of our algorithm. Based on their results, the population size is set to 100 for all experiments in this section. Additionally, the number of iterations of our algorithm is set from the range $[300..5000]$, based on the size of the graph.

For each problem size of our experiments, the edge density (α) is varied from the set $\{0.05, 0.1, 0.25, 0.5, 0.75, 1.0\}$ and the register density (β) is varied from the set $\{0.02, 0.04, 0.08, 0.1, 0.15, 0.2\}$. The spill costs are uniformly distributed in the range $[1..5]$. A set of 100 different interference graphs are generated with a fixed edge density value for each problem size; and each generated graph is tested with various number of registers by varying the register density. The total number of cases considered for each edge density is equal to 600.

Table 1. Average Results of Algorithms for Various Edge Density Values

		Test Cases			Unspilled Cases		Average Spilled Variables		Average Total Spill Costs	
n	α	HEA	OCH	Equal	HEA	OCH	HEA	OCH	HEA	OCH
100	0.05	132	6	462	500	473	4.28	5.32	10.04	11.53
	0.10	199	1	400	400	400	9.21	12.06	22.56	27.67
	0.25	400	0	200	301	200	19.86	25.70	51.69	62.43
	0.50	574	10	16	119	9	36.93	45.43	96.96	112.17
	0.75	589	9	2	0	0	56.07	64.18	147.58	162.29
	1.00	0	0	600	0	0	90.16	90.16	250.32	250.32
200	0.05	100	0	500	500	500	3.64	6.50	8.11	13.16
	0.10	199	1	400	436	400	12.03	17.43	30.06	39.75
	0.25	320	1	279	400	276	33.19	41.90	88.17	104.20
	0.50	499	1	100	200	100	63.62	80.52	171.68	201.45
	0.75	532	68	0	0	0	102.37	120.10	281.01	307.3
	1.00	0	0	600	0	0	180.33	180.33	513.00	513.00
300	0.05	100	0	500	500	500	2.65	5.95	5.40	10.47
	0.10	200	0	400	499	400	11.78	20.70	27.18	45.63
	0.25	300	0	300	400	300	42.38	56.68	109.28	138.77
	0.50	500	0	100	200	100	84.6	112.18	225.00	277.53
	0.75	516	84	0	73	0	144.28	175.75	395.72	436.45
	1.00	0	0	600	0	0	270.50	270.50	769.65	769.65
400	0.05	100	0	500	500	500	1.51	4.93	2.94	7.77
	0.10	179	0	421	500	421	12.23	23.64	26.42	49.15
	0.25	255	0	345	400	345	50.92	70.67	125.17	166.30
	0.50	465	0	135	200	135	105.2	142.31	271.05	341.66
	0.75	507	93	0	100	0	185.77	223.16	515.68	569.91
	1.00	0	0	600	0	0	360.67	360.67	1034.37	1034.37
500	0.05	100	0	500	501	500	0.77	3.58	1.52	5.24
	0.10	103	0	497	500	497	13.02	26.5	28.38	55.42
	0.25	200	0	400	400	400	60.03	84.42	150.65	202.27
	0.50	402	0	198	200	198	126.33	171.92	337.79	427.01
	0.75	412	188	0	100	0	227.93	271.67	600.91	650.30
	1.00	0	0	600	0	0	450.83	450.83	1269.79	1269.79

The first group of columns given in Table 1 is for the total number of test cases (out of 600) that our hybrid algorithm (HEA) produces better, worse or equal spill cost values with the optimistic coloring heuristic (OCH). Each row of the table also presents the number of unspilled test cases (out of 600), the average number of spilled variables, and the average total spill costs of the HEA and the OCH.

Our solution significantly outperforms the optimistic coloring algorithm with respect to the four metrics given in the table. When the edge density is very low, it will generate sparse graphs that may include disconnected nodes. A graph of this type is for the code where the live ranges do rarely overlap. When $\alpha = 0.05$, both algorithms provide the same results for 83% of test cases. Our algorithm outperforms OCH for 17% of the test cases, which is due to test cases when register density is equal to 0.02.

Table 2. Average Results of Algorithms for Various Register Density Values

n	β	Test Cases			Unspilled Cases		Average Spilled Variables		Average Total Spill Costs	
		HEA	OCH	Equal	HEA	OCH	HEA	OCH	HEA	OCH
100	0.02	482	16	102	0	0	68.75	72.46	191.09	198.13
	0.04	435	3	162	100	73	49.82	54.88	135.27	146.18
	0.08	297	3	300	201	200	33.57	38.73	88.26	97.68
	0.10	296	4	300	300	200	28.64	33.42	74.65	82.61
	0.15	200	0	500	319	300	20.09	24.65	51.18	58.27
	0.20	184	0	416	400	309	15.65	18.72	38.70	43.54
200	0.02	497	3	100	0	0	125.70	136.69	357.13	380.45
	0.04	398	2	200	136	100	91.53	101.32	255.74	274.54
	0.08	299	1	300	300	200	61.03	70.39	166.84	180.82
	0.10	156	65	379	300	276	52.62	60.33	143.18	152.83
	0.15	200	0	400	400	300	36.06	44.00	96.04	107.66
	0.20	100	0	500	400	400	28.24	34.05	73.08	82.55
300	0.02	500	0	100	0	0	177.1	196.45	499.86	543.62
	0.04	400	0	200	199	100	128.59	145.42	356.22	391.94
	0.08	294	6	300	300	200	85.88	99.67	234.68	255.10
	0.10	122	78	400	300	300	74.13	86.93	201.85	217.56
	0.15	200	0	400	400	300	50.42	61.49	134.20	151.62
	0.20	100	0	500	473	400	40.06	48.78	105.41	118.65
400	0.02	500	0	100	0	0	227.64	254.29	638.02	695.92
	0.04	379	0	221	200	121	165.11	188.85	454.45	504.96
	0.08	235	20	345	300	245	110.72	128.51	305.06	330.50
	0.10	127	73	400	300	300	94.82	112.18	261.78	282.22
	0.15	165	0	435	400	335	64.67	78.85	174.53	199.38
	0.20	100	0	500	500	400	53.33	62.68	141.78	156.17
500	0.02	500	0	100	1	0	279.15	310.73	779.29	848.48
	0.04	303	0	297	200	197	202.83	231.75	550.65	612.39
	0.08	102	98	400	300	300	136.18	157.34	367.24	394.21
	0.10	110	90	400	300	300	115.17	136.71	310.05	333.65
	0.15	102	0	498	400	398	78.91	96.21	208.28	235.31
	0.20	100	0	500	500	400	66.66	76.19	173.52	186.01

The number of cases that hybrid algorithm outperforms the other algorithm increases with an increase in edge density. When the generated graphs are fully connected (i.e., $\alpha = 1.0$), both algorithms give the same results for all test cases. Because of fully-connectiveness of the graphs, both algorithms spill all most all of the variables in the code, since the register density is in the range of $[0.02, ..., 0.20]$. Therefore, the number of test cases and average total cost values are equal for both HEA and OCH, when $\alpha = 1.0$. Our algorithm allocates registers to the variables without requiring spilling for more number of test cases than the optimum coloring heuristic. For both of the algorithms, an increase in edge density decreases the number of unspilled cases.

The performance of algorithms are examined for various register density values (see Table 2). Each row of this table gives the performance results of algorithms for a fixed register density, computed by averaging 600 randomly generated interference graphs. As in the previous tables, our algorithm outperforms the OCH with respect to all four performance metrics given. For each problem size, an increase in register density causes for more number of test cases to be allocated without spilling, which is true for both the HEA and the OCH. Additionally, the number of spilled variables and the total spill costs decrease with an increase in register density.

5 Conclusions

In this study, we introduced a hybrid evolutionary algorithm for register allocation problem in embedded devices. Our algorithm significantly outperformed a widely used register allocation heuristic with respect to all given parameters on solution quality. We also presented a highly specialized crossover and a domain specific local search technique. We are in the process of determining a test suit of benchmark routines, which will be used as part of the performance study.

References

1. P. Briggs, K. Cooper, K. Kennedy, L. Torczon, Coloring Heuristics for Register Allocation, Proc. of SIGPLAN 89 Conference on Programming Language Design and Implementation, pp. 275-284, Portland, Oregon, 1989.
2. P. Briggs, K. Cooper, L. Torczon, Improvements to Graph Coloring Register Allocation, ACM Transactions on Programming Languages and Systems, vol. 16, no. 3, pp. 428-455, 1994.
3. G. J. Chaitin, Register Allocation and Spilling via Graph Coloring, ACM SIGPLAN Conference on Programming Language Design and Implementation, pages 98-105, 1982.
4. Frederick Chow and John Hennessy, Register Allocation by Priority-Based Coloring, Proc. of SIGPLAN 84 Symposium on Compiler Construction, pp. 222-232, Montreal, Quebec, 1984.
5. M. G. Norman and P. Moscato, A Competitive and Cooperative Approach to Complex Combinatorial Search, Proc. of the 20th Informatics and Operations Research Meeting, Buenos Aires, August 1991.
6. B. Freisleben and P. Merz, A Genetic Local Search Algorithm for Solving Symmetric and Asymmetric Traveling Salesman Problems," Proc. of the IEEE International Conference on Evolutionary Computation, pp. 616-621, 1996
7. D. A. Fotakis, S. D. Likothanassis and S. K. Stefanakos, An Evolutionary Annealing Approach to Graph Coloring, Lecture Notes in Computer Science, vol. 2037, pp. 120-129, 2001.
8. D. Brelaz, New Methods to Color the Vertices of a Graph, Communications of the ACM, vol. 22 no. 4, pp. 251-256, 1979.
9. P. Galinier and J.-K. Hao, Hybrid Evolutionary Algorithms for Graph Coloring, Journal of Combinatorial Optimization, vol. 3, no. 4, 379-397, 1999
10. M. R. Garey and D. S. Johnson, Computers and Intractability: A Guide to the Theory of NP-Completeness, W. H. Freeman and Company, 1979.
11. R. Dorne and J. Hao, A New Genetic Local Search Algorithm for Graph Coloring, Springer Verlag, Lecture Notes in Computer Science, Vol. 1498, pp. 745-754, 1998.

Parallel Ant Systems for the Capacitated Vehicle Routing Problem

Karl F. Doerner[1], Richard F. Hartl[1], Guenter Kiechle[1], Maria Lucka[2], and Marc Reimann[1,3]

[1] Department of Management Science, University of Vienna,
Bruenner Strasse 72, A-1210 Vienna, Austria
{Karl.Doerner,Richard.Hartl,Guenter.Kiechle}@univie.ac.at
[2] Institute for Software Science, University of Vienna,
Liechtensteinstrasse 22, A-1090 Vienna, Austria
lucka@par.univie.ac.at
[3] Institute for Operations Research, Swiss Federal Institute of Technology Zurich,
Clausiusstrasse 47, CH-8092 Zurich, Switzerland
Marc.Reimann@ifor.math.ethz.ch

Abstract. In this paper we first provide a thorough performance comparison of the three main Ant Colony Optimization (ACO) paradigms for the Vehicle Routing Problem (VRP), namely the Rank based Ant System, the Max-Min Ant System and the Ant Colony System. Based on the results of this comparison we then implement a parallelization strategy to increase computational efficiency and study the effects of increasing the number of processors.

1 Introduction

The VRP is a well known combinatorial optimization problem, which has been extensively studied for the last 40 years. It involves the construction of a set of vehicle tours starting and ending at a single depot and satisfying the demands of a set of customers, where each customer is served by exactly one vehicle and neither vehicle capacities nor maximum tour lengths are violated.

The VRP belongs to the class of NP-hard problems (cf. [10]). Therefore no efficient exact solution methods are available, and the existing solution approaches are of heuristic nature. Recently the focus of research on this problem was on the use of meta-heuristics such as Tabu Search, Simulated Annealing and Ant Systems. An overview of meta- heuristic methods can be found [11].

In this paper we focus on solving the VRP with Ant Colony Optimization (ACO). Based on the observation of real ants' foraging behavior ACO was developed as a graph-based, iterative, constructive metaheuristic by Dorigo et al. (c.f. e. g. [8]). The main idea of ACO is that a population of computational ants repeatedly builds and improves solutions to a given instance of a combinatorial optimization problem. From one generation to the next a joint memory is updated that guides the work of the successive populations. The memory update is

J. Gottlieb and G.R. Raidl (Eds.): EvoCOP 2004, LNCS 3004, pp. 72–83, 2004.

based on the solutions found by the ants and more or less biased by their associated quality. Moreover, the reinforcement relates to the individual components of the solutions, e.g. the arcs in the Travelling Salesperson Problem (TSP).

There exist basically three different ACO paradigms which have evolved in parallel. These are the Rank based Ant System (ASrank) of Bullnheimer et al. [1], the Max-Min Ant System (MMAS) of Stuetzle and Hoos [18] and the Ant Colony System (ACS) proposed by Dorigo and Gambardella [7]. While there have been some efforts to get a 'fair' comparison and ranking of the three methods (c.f. [8]) the strength of the methods seem to really depend on the problem.

Independent of the actual choice concerning these ACO paradigms, the method, being a population based approach, seems to lend itself to parallelization on different levels. In fact recently some possible parallelization strategies for ACO have been proposed (c.f. e.g. [20]).

In this paper we first aim to clarify the relationship between the performances of the three paradigms for the VRP. We will then show the merit of a particular, rather simple, parallelization strategy for the paradigm deemed as 'best' from our initial analysis.

The remainder of this paper is organized as follows. In the next section we briefly review some approaches for the parallelization of optimization procedures. In this section we will also discuss some design issues arising in the context of parallelization. In Section 3 we describe the Savings based ACO and the differences in the pheromone management of the three paradigms, ASrank, MMAS and ACS. After that we turn to a description of our parallelization strategy in Section 4. Section 5 deals with the results of our computational tests before we conclude in Section 6.

2   Parallelization of Optimization Procedures

In recent years, some papers on the parallelization of algorithms for solving the capacitated vehicle routing problem have been published (e. g., [21], [15]). Jozefowiez et al. ([21]) developed a parallel Pareto genetic algorithm as well as a Pareto tabu search for a bi-objective VRP whereas Ralphs ([15]) developed a parallel exact procedure based on branch and cut for the problem at hand. A lot of applications were developed in the last years in the broader field of parallel computing in transportation (e. g., ([9])).

Parallelization strategies can be classified into ■ne-grained and coarse-grained strategies ([8]). In fine-grained parallelization strategies very few artificial ants are assigned to each processor and therefore frequent information exchange between the small sub-populations of ants (i.e. an information exchange between the processors) takes place. Such strategies are realized in [6], [12], [20], where the former two implementations require special hardware infrastructure, whereas for the ideas presented in [20] a cost effective PC-cluster is sufficient.

Coarse grained parallelization schemes run several subcolonies in parallel. The information exchange among these subcolonies is done at certain intervals or numbers of iterations (e. g. [2], [13]). The important questions in implementing different subpopulations are when, which and how information should be exchanged among the subcolonies (c.f. [13]). Stuetzle ([19]) studied the effect on solution quality when applying independent runs of the algorithm without communication in comparison to one longer run of the algorithm.

3   Savings Based ACO Algorithms for the VRP

All ACO algorithms mainly consist of the iteration of three steps:

- Generation of solutions by ants according to private and pheromone information
- Application of a local search to the ants' solutions
- Update of the pheromone information

The implementation of these three steps is described below. First, we will briefly outline the basic structure of the algorithm in terms of solution construction and local search as proposed in [17]. This basic structure is, with one modification in the decision rule of ACS, identical for all three ACO paradigms. Then we turn to the three paradigms and describe their mechanisms for pheromone management and in the case of ACS the modified decision rule.

3.1   Generation of Solutions

Solutions are constructed based on the well known Savings Algorithm due to Clarke and Wright, c.f. [4]. In this algorithm the initial solution consists of the assignment of each customer to a separate tour.
After that for each pair of customers i and j the following savings measure is calculated:

$$s_{ij} = d_{i0} + d_{0j} - d_{ij}, \tag{1}$$

where d_{ij} denotes the distance between locations i and j and the index 0 denotes the depot. Thus, the values s_{ij} contain the savings of combining two customers i and j on one tour as opposed to serving them on two different tours.

These values are then stored in a sorted list in decreasing order. In the iterative phase, customers or partial tours are combined by sequentially choosing feasible entries from this list. A combination is infeasible if it violates either the capacity or the tour length constraints.

The decision making about combining customers is based on a probabilistic rule taking into account both the above mentioned Savings values and the pheromone information. Let τ_{ij} denote the pheromone concentration on the arc connecting customers i and j telling us how good the combination of these two customers i and j was in previous iterations.

In each decision step of an ant, we consider the k best combinations still available, where k is a parameter of the algorithm which we will refer to as 'neighborhood' below. Let Ω_k denote the set of k neighbors, i.e. the k feasible combinations (i, j) yielding the largest savings, considered in a given decision step, then the decision rule is given by equation (2).

$$
\mathcal{P}_{ij} = \begin{cases} \dfrac{s_{ij}^{\beta} \tau_{ij}^{\alpha}}{\sum_{(h,l) \in \Omega_k} s_{hl}^{\beta} \tau_{hl}^{\alpha}} & \text{if } (i,j) \in \Omega_k \\ \\ 0 & \text{otherwise.} \end{cases}
\tag{2}
$$

In (2), $\mathcal{P}_{ij}$ is the probability of choosing to combine customers i and j on one tour, while α and β bias the relative influence of the pheromone trails and the savings values, respectively.

This algorithm results in a (sub-)optimal set of tours through all customers, once no more feasible savings values are available.

3.2 Local Search

A solution obtained by the above mentioned procedure can then be subjected to a local search in order to ensure local optimality. In our algorithms we sequentially apply the swap neighborhood (c.f. [16]) between tours to improve the clustering and the 2-opt algorithm (c.f. [5]) within tours to improve the routing.

3.3 Rank Based Ant System, Max-Min Ant System, and ACS

Let us now turn to the specific instantiations of the pheromone management in the three paradigms.

Rank based Ant System. The Rank based Ant System was proposed in [1] and its pheromone management centers around two concepts borrowed from Genetic Algorithms, namely ranking and elitism to deal with the trade-off between exploration and exploitation. In [17] this paradigm was used for solving the VRP. Thus, we will just briefly depict the pheromone update scheme here.

Let $0 \leq \rho \leq 1$ be the trail persistence and E the number of elitists. Then, the pheromone update scheme can formally be written as

$$
\tau_{ij} := \rho \tau_{ij} + \sum_{r=1}^{E-1} \Delta \tau_{ij}^{r} + E \Delta \tau_{ij}^{*}
\tag{3}
$$

First, the best solution found by the ants up to the current iteration is updated as if E ants had traversed it. The amount of pheromone laid by the elitists is $\Delta \tau_{ij}^{*} = 1/L^{*}$, where L^{*} is the objective value of the best solution found so far. Second, the $E - 1$ best ants of the current iteration are allowed to lay pheromone on the arcs they traversed. The quantity laid by these ants depends on their rank r as well as their solution quality L^{r}, such that the r-th best ant lays $\Delta \tau_{ij}^{r} = (E - r)/L^{r}$. Arcs belonging to neither of those solutions just face a pheromone decay at the rate $(1 - \rho)$, which constitutes the trail evaporation.

Max-Min Ant System. The Max-Min Ant System proposed in [18] is a different way to address exploration and exploitation issues.

First, only the global best solution found during the execution of the algorithm is reinforced. The modified pheromone update can be written as

$$\tau_{ij} := \rho\tau_{ij} + \Delta\tau_{ij}^{*}. \tag{4}$$

Again $0 \le \rho \le 1$ is the trail persistence and $\Delta\tau_{ij}^{*} = 1/L^{*}$, if an arc (i, j) belongs to the global best solution. Otherwise, $\Delta\tau_{ij}^{*} = 0$. Note, that while only the best solution is reinforced, all arcs are subject to evaporation.

As this leads to a stronger bias of the reinforcement towards good solutions and also causes the search to be more focused, it may lead to extensive exploitation and insufficient exploration. Thus additional components are added to the algorithm.

More specifically upper and lower bounds on the pheromone values are introduced to avoid stagnation caused by large differences between the pheromone values.

It can be shown that given the pheromone update shown above, the pheromone values are bounded by

$$\tau^{max} := \frac{1}{(1 - \rho) \cdot L^{*}}, \tag{5}$$

where L^{*} denotes the objective value of the optimal solution. Of course, this theoretical upper bound is of little use in the execution of the algorithm, as in general the value L^{*} of the optimal solution is unknown. Thus, L^{*} is replaced by L^{gb} to obtain the upper bound τ^{max} for the algorithm.

Note that this upper bound actually does not change anything with respect to the search strategy of the algorithm. However, it is used to determine the value for the lower bound on the pheromone values τ^{min}. By introducing a lower bound $\tau^{min} > 0$, the selection probability of rarely used solution elements is kept positive and thus diversity of solutions even after a large number of iterations is increased. By tuning the difference between the upper and lower bound the tradeoff between exploration and exploitation can be modelled. Note that due to the pheromone evaporation some values might decrease below τ^{min}. Thus, after the pheromone update such values are modified according to

$$\tau_{ij} := \max\{\tau^{min}; \tau_{ij}\}. \tag{6}$$

Finally, all pheromone values are initialized at the upper bound τ^{max}. This favors exploration in the early iterations of the algorithm as the pheromone only gradually evaporates, and reinforcement of good solution elements has a rather small impact. Over time some pheromone values will tend to the upper bound, while most will tend to the lower bound and the search turns from exploration to exploitation.

Ant Colony System. A different approach to deal with the problems associated with the tradeoff between exploration and exploitation called Ant Colony System (ACS) has been proposed in [7]. In that approach both the decision rule and the pheromone update differ from their counterparts in the other ACO algorithms.

The decision rule is based on an explicit modelling of exploitation. More specifically, a parameter q_0 is added that determines the probability to exploit the best option available in a decision step. The exact form of the decision rule is given by

$$
\mathcal{P}_{ij} = \begin{cases}
0 & \text{if } (i,j) \notin \Omega_k \\[2ex]
q_0 + (1-q_0)\dfrac{[s_{ij}]^\beta[\tau_{ij}]^\alpha}{\sum_{(h,l)\in\Omega_k}[s_{hl}]^\beta[\tau_{hl}]^\alpha} & \text{if } [s_{ij}]^\beta[\tau_{ij}]^\alpha = \max_{(h,l)\in\Omega_k}[s_{hl}]^\beta[\tau_{hl}]^\alpha \\[3ex]
(1-q_0)\dfrac{[s_{ij}]^\beta[\tau_{ij}]^\alpha}{\sum_{(h,l)\in\Omega_k}[s_{hl}]^\beta[\tau_{hl}]^\alpha} & \text{otherwise.}
\end{cases}
\tag{7}
$$

One advantage of this decision rule, besides the increased exploitation of the best solution, is the increase in speed, as the time consuming roulette wheel selection is only applied with a probability $1 - q_0$.

As stated above, not only the decision rule in ACS is different to the one used in the other ACO paradigms, but also the pheromone update is changed.

First, as in the Max-Min Ant System described above only the global best solution found so far is reinforced. However, to avoid extensive exploitation of the best solution, global evaporation is also restricted to the elements of this global best solution, while all other elements are left unchanged. In this respect the pheromone update differs from the update used in the Max-Min Ant System, where evaporation is applied to all arcs. Formally this can be written as

$$
\tau_{ij} := \rho\tau_{ij} + (1-\rho)\Delta\tau_{ij}^* \quad \text{if } (i,j) \in L^*.
\tag{8}
$$

Further the updated pheromone values in this approach are a convex combination of the pheromone values before the update and the update quantity. This mechanism adjusts the speed at which pheromone values tend to their upper bound and thus increases exploration.

Second, exploration is forced by an added mechanism, namely local evaporation. Each time a decision is taken, the corresponding solution element is made less attractive. Thus, even for a very high level of q_0 a decision can not be taken over and over again. Rather at some point it will not be the most attractive option and alternative choices will be made. More precisely, local evaporation is done as follows:

$$
\tau_{ij} := \rho\tau_{ij} + (1-\rho)\tau^{min}.
\tag{9}
$$

Note that as in the Max-Min Ant System pheromone values are bounded below by the value τ^{min}.

4   Parallelization Strategy for Our Savings Based ACO

A common procedure for implementing parallel programs (especially on distributed memory machines) is to employ low level message passing systems such as MPI ([14]) in conjunction with a standard sequential language such as Fortran or C. With this approach explicit communication calls for non local data access have to be inserted into the sequential program.

Note that our goal in parallelization of the ACO algorithm is to improve the execution time of the algorithm without altering its behavior. Possible improvements of ACO solution quality through the exploitation of parallelism will be analyzed in another project and are not discussed in this article.

Figure 1 shows the basic setup of our parallel implementation. Generally speaking, we parallelize the ACO on the basis of individual ants.

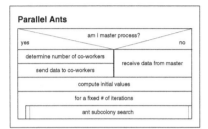
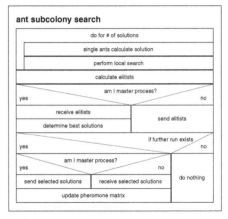

Fig. 1. Communication structure of the processes

We have enriched our C implementation with MPI statements to exploit the parallelism in the algorithm in the following way: The input data are read by master process and distributed to all coworker processes. The different processes independently initialize the algorithm and compute the distance matrix. Each process constructs solutions, where the number of constructed solutions by one process is computed by the number of ants divided by the number of processes. As soon as the solutions of all processes are available the n-best solutions are searched and broadcasted to the master process. If the number of ants is smaller then the number of the n-best ants then the solutions of all the ants are sent to the master process. The reduction operation for the determination of the n-best ants is executed - then the solutions of the n-best ants are broadcasted from master process to the different coworker processes and the pheromone matrices are updated in each process.

5   Computational Analysis

The computational analysis is based on 14 classic benchmark instances for the VRP (c.f. [3])). These instances range from 50 to 199 customers. The set of instances is comprised of both, problems where all customers are randomly located in the plane and problems where customer locations are arranged in several clusters. Further, some instances are just capacity constrained while others also feature restrictions on the maximum tour length.

5.1   Comparison of ASrank, MMAS, and ACS for the VRP

Our ACO algorithms were implemented in C and the comparison of the three ACO paradigms is based on experiments performed on a Pentium 4 with 2.6 GHz with double precision floating point arithmetic. We have tuned each algorithm separately with respect to the following parameters: $\alpha, \beta, \rho, E, k, q_0, \tau_{min}$, the number of ants and the number of iterations.

The most important results of this tuning are summarized below. Table 1 presents for several algorithm settings the average deviation of our best, average and worst results over 5 runs from best known solutions (RPD) for all instances, together with the average time in minutes needed to find the best solution.

Table 1. Performance comparison of 3 ACO variants for the VRP

	ASrank		MMAS			ACS			
α	5	5	1	1	1	1	1	1	1
β	5	5	2	5	2	2	2	2	2
ρ	0.95	0.975	0.98	0.98	0.98	0.9	0.9	0.9	0.95
E	4	4	1	1	1	1	1	1	1
k	$n/4$	$n/4$	$n/4$	$n/4$	$n/4$	$n/2$	$n/4$	$n/4$	$n/4$
q_0	——	——	——	——	——	0.9	0.9	0.9	0.9
τ_{min}	0	0	0.0001	0.0001	0.0001	0.0001	0.0001	0.0001	0.0001
# of ants	n	n	n	n	$n/2$	10	10	10	10
# of iterations	$2n$	$2n$	$5n$	$5n$	$10n$	$15n$	$15n$	$30n$	$30n$
RPD best run	0.61	0.5	0.64	0.57	0.67	0.9	0.67	0.68	0.54
RPD avg. run	1.14	0.97	0.89	0.8	0.95	1.61	1.52	1.31	1.22
RPD worst run	1.72	1.41	1.15	1.02	1.29	2.56	2.46	2.25	2.12
t_{best} (min.)	0.93	1.75	4.89	4.12	3.55	2.07	1.04	2.02	2.25

Let us first look at each of the three paradigms separately. For ASrank, Table 1 shows that a significant improvement could be obtained by changing the trail persistence from $\rho = 0.95$ to $\rho = 0.975$. However, this improvement was achieved at the cost of an almost twofold increase in computation times.

For MMAS we show three of the best parameter settings. First, we can see that increasing β leads to both an improvement in solution quality and computation times. Second, we observe that reducing the number of ants while increasing the number of iterations - thus keeping the total number of solutions computed constant - leads to a further improvement in computation time, i.e. earlier convergence however at the cost of a clear deterioration in solution quality.

Finally, for ACS Table 1 reveals the following findings. First reducing the size of the neighborhood from $k = n/2$ to $k = n/4$ leads to a twofold increase in computation times and also improves upon the solution quality. Second, doubling the number of iterations, yields another improvement in the average and worst RPD however at the cost of twice as much computational effort. Third, increasing rho leads to a further improvement in solution quality at a moderate increase in computation times.

Let us now turn to a brief discussion of the effects some other parameters had on the performance of the three paradigms. We will do this for each of the three paradigms separately. When we talk about changes in the parameters we will always refer to these changes relative to the settings presented in Table 1.

ASrank. For ASrank it turned out that the settings proposed in [1] and [17] are quite good. The following parameter changes had the largest impact on either solution quality or computational times.

- size of neighborhood k: Increasing the size of the neighborhood k leads to insignificant improvements in solution quality at the cost of strongly increased computation times.
- number of ants: Reducing the number of ants leads to a strong tendency towards premature convergence and thus a high variance in the results.

MMAS. For the MMAS the following parameters turned out to be crucial in our tests.

- trail persistance ρ: Increasing ρ leads to a huge increase in computation times particularly for larger instances. Reducing ρ leads to a strong tendency towards premature convergence and thus a high variance in the results.
- pheromone weight α: Increasing α, while reducing the computational burden significantly lead to a huge deterioration in solution quality through premature convergence.

ACS. In ACS, changes in the following parameters had the most severe effects on the performance of the algorithm.

- number of ants: Increasing the number of ants, significantly increases computational times without really improving upon the solution quality. In our runs an increase in the number of iterations was always preferable over an increase in the number of ants.
- lower pheromone limit τ_0: Reducing τ_0 leads to a strong deterioration of solution quality, while computation times do not change.
- level of exploitation q_0: Changing the level of exploitation in both directions leads to a significant negative effect on solution quality. In the case of a reduction in the level of exploitation this is accompanied by an increase in computational time.

Comparing the three paradigms with each other, the following results can be deduced from Table 1. Particularly with respect to average and worst case performance ASrank and MMAS seem to outperform ACS. While MMAS seems to be slightly better than ASrank when measured in terms of solution quality, the results show that ASrank obtains good solutions much faster than MMAS. This is clearly due to the broad search in the beginning of an MMAS run that is needed to avoid premature convergence. On the other hand, ACS is very fast but the solution quality obtained shows a high variance. Summarizing, it seems that MMAS suffers from slow convergence due to its exploratory phase in the beginning, while ACS follows a too narrow search trajectory. On the other hand, ASrank seems to best find the balance between computation time and solution quality a feature particularly important for large problems.

Thus, we will use ASrank for the analysis of efficiency aspects associated with parallelization. As pointed out in Section 4, our goal in parallelization of the ACO algorithm is to improve the execution time of the algorithm without altering its behavior.

5.2   Efficiency Effects of Parallelization

The experiments with the parallelized version of the Savings based ASrank were run on a state-of-the art medium-sized Beowulf cluster comprised of 8 quad-board PCs equipped with Pentium III processors and connected via Fast-Ethernet. For analyzing the effectiveness of our parallel algorithm we measured the speedup as well as the efficiency obtained on varying numbers of processors. Speedup is defined as

$$S_p = T_{seq}/T_p \tag{10}$$

where S_p is the speedup obtained on p processors, T_{seq} is the measured sequential execution time and T_p is the parallel execution time on p processors. Efficiency is defined as

$$E_p = S_p/p \tag{11}$$

For measuring the sequential execution time we compiled the program without the MPI statements with native C compilers available on any architecture.

Speedup and efficiency obtained with different numbers of processors are shown in Figure 2. More precisely, we present average results over 2 of the benchmark problem instances, namely those with 150 customers. We obtained qualitatively identical results for the other instances.

Figure 2a clearly shows the increasing speedup as the number of processors increases from 1 to 32. However, from Figure 2b it is obvious that efficiency decreases at an increasing rate as the number of processors goes up. Thus, overheads such as communication costs jeopardize the gains from executing the algorithm on parallel machines. Still it seems that using up to 8 processors is reasonable as a sixfold speed up is obtained with an efficiency of still more than 70% and even with 32 processors the gains from parallel execution of the program are larger than the communication costs.

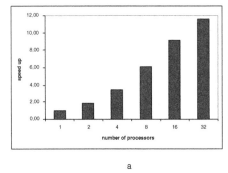 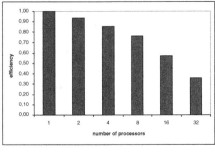

a   b

Fig. 2. Speed-up and Efficiency

Note, that to our best knowledge this is the first paper to show the speed-up and efficiency effects of parallelization for up to 32 processors. Most of the existing parallelization papers focus on the solution quality and competitiveness effects of their parallel implementation.

6   Conclusion

In this paper we have first provided a thorough performance comparison of the three main ACO paradigms for the VRP. Based on the results of this comparison we have then shown the merit, in terms of the increase in computational efficiency, of a parallelization strategy. We are currently extending our endeavors concerning parallel implementations of ACO algorithms to more sophisticated models of parallelization. The aim is to come up with more intelligent strategies that will lead to either better efficiency or better effectiveness (or possibly both). Also, the speed-up obtainable through parallelization should be highly important for solving larger problem instances in real time.

Acknowledgments. We wish to thank B. Mermer for running the computational experiments. We are also grateful to G. Haring, the dean of the School of Business and Informatics at the University of Vienna for financial support, and S. Benkner for providing us with know-how concerning parallel implementations. Finally, we acknowledge the valuable comments of three anonymous referees.

References

1. Bullnheimer, B., Hartl, R. F. and Strauss, C.: "A new rank based version of the ant system: a computational study". *Central European Journal of Operations Research* **7**(1) (1999), p. 25–38.
2. Bullnheimer, B., Kotsis, G. and Strauss, C.: "Parallelization Strategies for the Ant System". In: R. Leone et al. (eds.): *High Performance Algorithms and Software in Nonlinear Optimization*, Kluwer, Dordrecht (1998), p. 87–100.

3. Christofides, N., Mingozzi, A. and Toth, P.: "The vehicle routing problem". In: Christofides, N. et al. (eds.): *Combinatorial Optimization*. Wiley, Chicester (1979).

4. Clarke, G. and Wright, J. W.: "Scheduling of vehicles from a central depot to a number of delivery points". *Operations Research* **12** (1964), p. 568–581.

5. Croes, G. A.: "A method for solving Traveling Salesman Problems". *Operations Research* **6** (1958), p. 791–801.

6. Delisle, P., Krajecki, M., Gravel, M. and Gagne, C.: "Parallel implementation of an ant colony optimization metaheuristic with OpenMP". *Proceedings of the 3rd European Workshop on OpenMP (EWOMP'01)*, September 8–12, 2001, Barcelona, Spain.

7. Dorigo, M. and Gambardella, L. M.: "Ant Colony System: A cooperative learning approach to the Travelling Salesman Problem". *IEEE Transactions on Evolutionary Computation* **1**(1) (1997), p. 53–66.

8. Dorigo, M. and Stuetzle, T.: "The Ant Colony Optimization Metaheuristic: Algorithms, Applications, and Advances". In: Glover, F. and Kochenberger, G. A. (eds.): *Handbook of Metaheuristics*, Kluwer, p. 251–285.

9. Florian, M. and Gendreau, M.: "Applications of parallel computing in transportation". *Parallel Computing* 27 (2001), p. 1521–1522.

10. Garey, M. R. and Johnson, D. S.: *Computers and Intractability: A Guide to the Theory of NP Completeness.* W. H. Freeman & Co., New York (1979).

11. Gendreau, M., Laporte, G. and Potvin, Y.: "Metaheuristics for the vehicle routing problem". GERAD Report G-98-52, University of Montreal, Canada (1999).

12. Merkle, D., and Middendorf, M., "Fast Ant Colony Optimization on Runtime Reconfigurable Processor Arrays". *Genetic Programming and Evolvable Machines*, **3** (4) (2002), p. 345–361.

13. Middendorf, M., Reischle, F. and Schmeck, H., "Multi Colony Ant Algorithms". *Journal of Heuristics* **8** (2002), p. 305–320.

14. MPI: "MPI: A message-passing interface standard version 1.1" MPI Forum (1995).

15. Ralphs, T. K., "Parallel branch and cut for capacitated vehicle routing", *Parallel Computing* **29** (2003), p. 607–629.

16. Osman, I. H.: "Metastrategy simulated annealing and tabu search algorithms for the vehicle routing problem". *Annals of Operations Research* **41** (1993), p. 421–451.

17. Reimann, M., Stummer, M. and Doerner, K. (2002): "A Savings based Ant System for the Vehicle Routing Problem". In: Langdon, W. B. et al. (eds.): *Proceedings of GECCO 2002*, Morgan Kaufmann, San Francisco, p. 1317–1325.

18. Stuetzle, T. and Hoos, H.: "The Max-Min Ant System and Local Search for Combinatorial Optimization Problems". In: Voss, S. et al. (eds.): *Meta-Heuristics: Advances and Trends in Local Search Paradigms for Optimization*, Kluwer (1998).

19. Stuetzle, T.: "Parallelization Strategies for Ant Colony Optimization". In: Eiben, A. E. et al. (eds): *PPSN-V*, Amsterdam, Springer, LNCS 1498, (1998), p. 722–731.

20. Talbi, E.-G., Roux, O., Fonlupt, C. and Robillard, D.: "Parallel Ant Colonies for the quadratic assignment problem". *Future Generation Computer Systems*, **17** (2001), p. 441–449.

21. Jozefowiez, N., Semet, F. and Talbi, E.-G.: "Parallel and Hybrid Models for Multiobjective Optimization: Application to the Vehicle Routing Problem". In: Geuervos, M. et al. (eds.), *PPSN VII*, LNCS 2439, (2002), p. 271–280.

A Hierarchical Social Metaheuristic
for the Max-Cut Problem

Abraham Duarte[1], Felipe Fernández[2], Ángel Sánchez[1], and Antonio Sanz[1]

[1] ESCET-URJC, Campus de Móstoles, 28933, Madrid, Spain
{a.duarte, an.sanchez, a.sanz}@escet.urjc.es
[2] Dept. Tecnología Fotónica, FI-UPM, Campus de Montegancedo, 28860, Madrid, Spain
Felipe.Fernandez@es.bosch.com

Abstract. This paper introduces a new social metaheuristic for the Max-Cut problem applied to a weighted undirected graph. This problem consists in finding a partition of the nodes into two subsets, such that the sum of the weights of the edges having endpoints in different subsets is maximized. This NP-hard problem for non planar graphs has several application areas such as VLSI and ASICs CAD. The hierarchical social (HS) metaheuristic here proposed for solving the referred problem is tested and compared with other two metaheuristics: a greedy randomized adaptive search procedure (GRASP) and an hybrid memetic heuristic that combines a genetic algorithm (GA) and a local search. The computational results on a set of standard test problems show the suitability of the approach.

1 Introduction

An important graph bipartition problem is the Max-Cut problem defined for a weighted undirected graph $S = (V, E, W)$, where V is the ordered set of vertices or nodes, E is the ordered set of undirected arcs or edges and W is the ordered set of weights associated with each edge of the graph. This Max-Cut optimization problem consists in finding a partition of the set V into two disjoint subsets (C, C') such that the sum of the weights of edges with endpoints in different subsets is maximized. Every partition of vertices V into C and C' is usually called a cut or cutset and the sum of the weights of the edges is called the weight of the cut.

For the considered Max-Cut optimization problem, the cut value or weight of the cut given by

$$w(C, C') = \sum_{v \in C, u \in C'} w_{vu} \tag{1}$$

is maximized. In [9] is demonstrated that the decision version of Max-Cut is NP-Complete. This way, we need to use approximate algorithms for finding the solution in a reasonable time. Notice that for planar graphs exacts algorithms exist which solve the Max-Cut problem in polynomial time [10].

In this paper we propose a new hierarchical social (HS) metaheuristic for finding an approximate solution to Max-Cut problem. Additionally, two standard metaheruristics:

J. Gottlieb and G.R. Raidl (Eds.): EvoCOP 2004, LNCS 3004, pp. 84–94, 2004.

GRASP and memetic algorithms (MA) are analyzed and compared with the HS algorithm.

The new introduced metaheuristic, hierarchical social algorithms [3,4], is inspired in the behaviour of some hierarchical societies. HS algorithms make use of local search heuristics with a group-population-based strategy. Their basic structure is a graph where a dynamical domain partition is performed. In each region of the considered partition a hierarchical group is settled. Each group consists of a core that determines the value of the corresponding group objective function and a periphery that defines the local search region of the involved group. Through a competitive group population strategy, the set of groups is optimized, the number of groups successively decreases and an approximate optimal solution is found.

A second metaheuristic used to compare with HS algorithms is a Greedy Randomized Adaptive Search Procedure (GRASP) [5,15,16], based on a construction phase and a local optimization phase.

A third metaheuristic also used to compare the experimental results, is based on a genetic algorithm [7,12,17] with an additional local search based on the problem domain knowledge (memetic algorithm [13,14]), which allows a remarkable improvement of the solution obtained.

2   Review of Max-Cut Problem Solution

The Max-Cut problem can be formulated as an integer quadratic program [18]. This program can not be efficiently solved because, in the general case, it is a NP-complete problem [9]. For planar graphs exact polynomial time algorithms exist [10]. Several continuous and semidefinite relaxations have been proposed in order to achieve a good solution in a reasonable running time. In [11,18] is described the semidefinite relaxation of the Max-Cut problem. Goemans et al. showed in [6] a randomized algorithm that guarantees a 0.878-approximation to the optimum and in addition an upper bound on the optimum. Based on this work, other approximation algorithms have been proposed [2].

There are other approaches to the Max-Cut problem relaxation [2]. Maybe, one of the most interesting is the 2-rank relaxation proposed in [1] that in practice produces better solutions than the mentioned randomized algorithm [6].

3   Hierarchical Social Algorithms for the Max-Cut Problem

This section presents the general features of a new metaheuristic called hierarchical social (HS) algorithms. In order to get a more general description of this metaheuristic, the reader is referred to [3][4].

HS algorithms are inspired in the hierarchical social behaviour observed in a great diversity of human organizations and biological systems.

This metaheuristic have been successfully applied to several problems such as critical circuit computation [4] and DFG scheduling with unlimited resources [3].

The key idea of HS algorithms is a simultaneous optimization of a set of feasible solutions. Each group of a society contains a feasible solution and these groups are initially random distributed through the solution space. By means of an evolution strategy, where each group tries to improve itself or competes with the neighbour groups,
better solutions are obtained through the corresponding social evolution.

In this social evolution, the groups with lower quality tend to disappear. As a result,
the rest of the group objective functions are optimized. The process usually ends with
only one group that contains the best solution found.

3.1   Metaheuristic Structure

For the Max-Cut problem, the feasible society is modelled by the specified undirected
weighted graph $S=(V,E,W)$ also called feasible society graph. The set of individuals
are modelled by nodes V of the specified graph, and the set of feasible relations are
modelled by arcs A of the specified graph. Figure 1.a shows an example of the feasible
society graph for a particular Max-Cut problem.

The state of a society is modelled by a hierarchical policy graph. This hierarchical
policy graph also specifies a society partition composed by a disjoint set of groups
$\Pi=\{g_1, g_2,...,g_g\}$, where each individual or node is assigned to a group. Each group
$g_i \subset \Pi$ is composed by a set of individuals and active relations, which are constrained
by the feasible society. The individuals of all groups cover the individuals of the
whole society.

The specification of the hierarchical policy graph is problem dependent. The initial
society partition determines an arbitrary number of groups and assigns individuals to
groups. Figure 1.b shows a society partition example formed by two groups.

A society partition can be classified into *single-group or monopoly partition*, in
which there is only one group, and *multiple-group or competition partition*, in which
there are more than one group. Obviously in example shown in Figure 1.b, the partition shown is a competition partition.

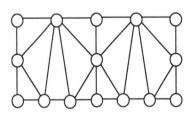 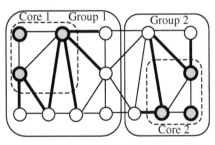

Fig. 1. (a) Feasible society graph. (b) Society partition and groups partition

Each individual of a society has its individual objective function $f1$. Each group
has its group objective function $f2$ that is shared by all individuals of a group. Furthermore each group g_i is divided into two disjoint parts: core and periphery. The core
determines the value of the corresponding group objective function $f2$ and the periphery defines the local search region of the involved group.

In the Max-Cut problem framework, the set of nodes V_i of each group g_i is divided into two disjoint parts: $g_i=(C_i, C_i')$ where C_i represents the *core* or subgroup of nodes that belongs to the considered cutset and C_i' is the complementary subgroup of nodes or *periphery*. The core edges are the arcs that have their endpoints in C_i and C_i'. Figure 1.b also shows an example of cores for the considered previous partition. The core edges of each group are shown in this figure by thick lines.

For each group $g_i = (C_i, C_i')$, the group objective function $f2(i)$ is given by the function $cut(i)$:

$$f2(i) = cut(i) \quad \text{where} \quad cut(i) = \sum_{v \in C_i, u \in C_i'} w_{vu} \qquad (2)$$

$$\forall v \in g_i \quad f2(v,i) = f2(i) = cut(i)$$

where the weights w_{vu} are supposed to be null if the corresponding edges do not belong to the specified graph. Obviously, this value is determined by the core edges, because is the sum of the edges weights which have one endpoint in the core C_i, and the other endpoint in the periphery C_i' of the corresponding group g_i.

For each individual or node v, the individual objective function $f1(v,i)$ relative to each group $g_i = (C_i, C_i')$ is given by the function:

$$f1(v,i) = max(\sigma(v,i), \sigma'(v,i)) \quad \text{where} \qquad (3)$$

$$\sigma(v,i) = \sum_{u \in C_i} w_{vu} \quad \text{and} \quad \sigma'(v,i) = \sum_{u \in C_i'} w_{vu}$$

The HS algorithms here considered, try to optimize one of their objective functions ($f1$ or $f2$) depending on the operation phase. During an autonomous or winner phase, each group i tries to improve independently the group objective function $f2$. During a loser phase, each individual tries to improve the individual objective function $f1$, the original groups cohesion disappeared and the graph partition is modified in order to optimize the corresponding individual objective function. The strategy followed in a loser phase could be considered inspired in Adams Smith's "invisible hand" economic society mechanism described in his book "An Inquiry into the Nature and Causes of the Wealth of Nations".

3.2   Metaheuristic Process

The algorithm starts from a random set of feasible solutions i.e. it obtains an initial partition that determines the corresponding groups structure. Additionally for each group, an initial random cutset is derived.

The groups are successively transformed through a set of competition phases based on local search strategies. For each group, there are two main strategies: *winner or autonomous strategy* and *loser strategy*. The groups named *winner groups*, which apply the winner strategy, are that which have the higher group objective function $f2$. The rest of groups apply loser strategy and are named *loser groups*. During optional *autonomous phases* in between competition phases, all groups behave like winner groups. These optional autonomous phases improve the capability of the search procedure.

The winner strategy can be considered as a local search procedure in which the quality of the solution contained in each group is improved by autonomously working with the individuals that belong to each group and the relations among these individuals.

The loser strategy is oriented to let the exchange of individuals among groups. In this way the groups with inferior quality (lower group objective function $f2$) tend to disappear, because their individuals move from these groups to another groups that grant their maximum individual promotion (highest individual objective function $f1$).

Winner and loser strategies are the basic search tools of HS algorithms. Individuals of loser groups, which have lower group objective functions, change of group during a loser strategy, in order to improve their individual objective function $f1$. This way, the loser groups tend to disappear in the corresponding evolution process. Individuals of winner groups, which have highest group objective functions, can move from the core to periphery or inversely, in order to improve the group objective function $f2$. These strategies produce dynamical groups populations, where group annexations and extinctions are possible.

3.3   High-Level Pseudo-Code

A general high-level description of an HS metaheuristic is shown in Figure 2, where the main search procedures are: Winner Strategy and Loser Strategy.

In the Max-Cut problem, the Winner Strategy (Figure 3) is oriented to improve the cut value of a group. For a given group g_i, a nodes exchange, between core and periphery, allows to optimize the group objective function. The exchange is accomplished if there is an improvement of the cut weight or objective function $f2(i)$ of the corresponding group g_i. This procedure is based on the local search procedure presented in [15,16]. In our implementation, we also allow to move the nodes between C_i and C_i' in parallel and simultaneous way on a single iteration. The only restriction is the following: if one node u is gone out from C_i (or C_i'), none of its adjacent nodes can change their position in the same iteration. This restriction avoids cycling in the corresponding procedure.

```
Procedure Hierarchical_Social_Algorithm(S)
Var
  S=(V,E,W)=({v}{e}{w}):Initial_Society_graph;
  G={gi}:Groups_structure;
  F1={f1}:Individuals_objective_function_structure;
  F2={f2}:Groups_objective_function_structure;
  i,k:1..Number_of_groups /*Group indices*/
Begin /* Begin social evolution */
  G=Get_initial_random_partition_and_groups_structure();
  Repeat   /* group evolution*/
   Compute_F1()_and_F2(); /*Objective functions*/
   For each gi in G
     If f2(i)=max{f2(k)∀k} Or Autonomous_phase Then
       Winner_Strategy(i)
     Else
       Loser_Strategy(i);
   End For
   Update_groups_structure(G);
  Until termination_condition_met;    /*End of social evolution*/
  Return(G); /* Approximate optimal solution */
End Hierarchical_Social_Algorithm
```

Fig. 2. High-level pseudo-code for Hierarchical Social Algorithms

For each vertex v and each group $g_i = (C_i, C_i')$ the following function σ and σ' are considered:

$$\sigma(v,i) = \sum_{u \in Ci} w_{vu} \quad \sigma'(v,i) = \sum_{u \in Ci'} w_{vu} \tag{4}$$

These evaluation functions are used in the pseudo-code of Figure 3.

The Loser Strategy allows an individual to move to another group. The individuals belonging to groups with lower cut value $f2(i)$ can change their group in order to increase the individual objective function $f1(v,i)$. Each node searches for the group j that gives the maximum improvement in its individual objective function. Figure 4 shows the pseudo-code of Loser Strategy considering the functions σ and σ' defined in (4).

```
Procedure Winner_Strategy(i)/* for Max_Cut problem*/
Var gi=(Ci,Ci´):Cutset_structure_of_considered_group
Begin
For v = 1 to Unmber_of_nodes_of_considered_group
   If v∈Ci and σ(v,i)>σ'(v,i) Then Ci´=Ci'∪{v}; Ci=Ci\{v}; /*v→Ci´*/
   If v∈Ci´and σ(v,i)<σ'(v,i) Then Ci'= Ci'\{v}; Ci=Ci∪{v};/*v→Ci */
End For
End Winner_Strategy
```

Fig. 3. Pseudo-code of Winner Strategy

```
Procedure Loser_Strategy(i)/* for Max_Cut problem*/
Var
 gi=(Ci,Ci´):Cutset_structure_of_considered_group
 G={gi}={(Ci,Ci´)}:Cutset_structure_of_groups
 i,j,k:1..Number_of_groups /*Group indices*/
Begin
For v = 1 to Number_of_nodes_of_considered_group
   j= arg max {max(σ(v,k),σ'(v,k)),∀k} /*j=host group*/
   If (v∈Ci) Then Ci=Ci\{v} Else Ci´=Ci´\{v}; /*Remove v from gi*/
   If σ(v,j)>σ'(v,j) Then Cj'=Cj'∪{v} Else Cj=Cj ∪{v}; /*Add v to gj*/
End For
End Loser_Strategy
```

Fig. 4. Pseudo-code of Loser Strategy.

4   GRASP for MAX-CUT Problem

This section briefly summarizes the use of a Greedy Randomized Adaptive Search Procedure (GRASP) for solving the Max-Cut problem. For a more detailed description the reader is referred to references [5,15,16]. GRASP is a multi-start iterative method where in each iteration there are two phases, construction phase and local search phase. In the first phase, a greedy feasible solution is constructed and in the second phase, starting from this solution, a local optimal solution is derived. Figure 5 shows a high level pseudo-code of this metaheuristic, assuming that the objective function f is the same that the group objective function ($f2$) introduced in (2).

The construction phase makes of an adaptive greedy function that uses a restricted candidates list (RCL) and a probabilistic selection criterion [15,16]. In the Max-Cut problem, for each node $v \notin C \cup C'$, the following greedy function $f(v)$ is considered:

$$\sigma(v) = \sum_{u \in C} w_{vu} \quad \sigma'(v) = \sum_{u \in C'} w_{vu} \tag{5}$$

$$f(v) = max\{\sigma(v), \sigma'(v)\} \qquad \forall v \notin C \cup C' \tag{6}$$

This greedy function is used to create an RCL based on the specified cut-off value α. Later on, a random procedure successively selects vertices from this RCL in order to obtain an initial feasible solution.

```
Procedure GRASP(S,Max_number_of_iterations)
Var g:Solution_structure;
Begin
 For i=1 to Max_number_of_iterations   /*Multi-start*/
 g= Build_a_greedy_randomized_solution(S);
 g= Local_Search(g);
 If i = 1 Then g* = g; /*Initial solution*/
 Else f(g)>f(g*) Then g* = g; /*Update solution*/
 End For
 Return(g*); /* Approximate optimal solution */
End GRASP;
```

Fig. 5. Pseudo-code of a generic GRASP

The local search phase is later applied, which is also based on the previous definitions of σ and σ', taking into account that all nodes have already been assigned to C or C'. If a node $v \in C$ and $\sigma(v) > \sigma'(v)$ then it should be changed from C to C': $C'=C'\cup\{v\}$ and $C= C\backslash\{v\}$, and conversely. This second phase stops when all possible movements have been computed and no improvement was found.

5   Memetic Algorithms for the MAX-CUT Problem

This section briefly summarizes the use of genetic algorithms for solving the Max-Cut problem. For a more detailed description the reader is referred to references [7,12,14,17].
The introduction of an additional local search phase on a genetic algorithm produces a memetic algorithm. This improves the quality of the solutions obtained. In Figure 6 is shown the high level pseudocode of a memetic algorithm.

```
Procedure_Memetic_Algorithm(S, Number_generations)
 Initialize_random_population(S);
 Local_Search(); /*Local search used in GRASP */
 For gen =1 to Number_generations
   Select_parents(gen);
   Produce_offspring_by_fixed_crossover_of_parents(gen);
   Local_Search(gen); /*Local search used in GRASP */
   Produce_mutation(gen);
   Evaluate_the_offspring(gen);
   Replacement_process(gen);
 End For
End Memetic_Algorithm
```

Fig. 6. High-level general pseudo-code of used memetic algorithm.

In order to use memetic algorithms for solving the Max-Cut problem, we need to code each feasible solution. Let $V = \{1,\ldots,n\}$ the nodes set of a given graph. The cuts of this graph can be coded by a Boolean n-vector $I = (i_1,\ldots,i_n)$ such that the value of each $i_u \in \{0,1\}$ is given by the characteristic function: $i_u = \{(u \in C) \rightarrow 1; (u \in C') \rightarrow 0\}$:

In the evaluation step, the selection method used is the fitness roulette-wheel selection [12], which favours individuals with high fitness without suppressing the chance of selection of individuals with low fitness, avoiding premature convergence of the population.

This way, a subset on individuals is selected, introduced in the new population and recombined with a probability p_r. In this paper we consider the crossover method, called fixed crossover [2], which is dependent on the structure of the crossed individuals. In general, the use of structural information is an advantage providing more quality descendants.

The fixed crossover function considered $f: \{1,0\} \times \{1,0\} \rightarrow \{1,0\}$ is specified by the following random Boolean function:

$$f((\alpha,\beta)) = \{(0,0) \rightarrow 0; (0,1) \rightarrow rand01; (1,0) \rightarrow rand01; (1,1) \rightarrow 1\} \qquad (7)$$

where rand01 is a random Boolean value. In this way, if both parents are in the same subset, the offspring node lies in this subset. Otherwise, the node is randomly assigned to one of the subsets. With this crossover function, each bit i_u of new offspring is given by:

$$i_u = f(father(i_u), mother(i_u)) \qquad \forall u \in V \qquad (8)$$

To ends up with the evolution cycle, new individuals are subject to mutation (a random change of a node from C to C' or inversely) with probability p_m.

6   Experimental Results

In this section, we describe the experimental results for the three considered metaheuristics. The computational experiments were evaluated in an Intel Pentium 4 processor at 1.7 GHz, with 256 MB of RAM. All algorithms were coded in C++, without optimization and for the same programmer in order to have more comparable results.

The main objective of this section consists in comparing different types of evolutionary approaches. In some sense, the memetic algorithms are in one extreme (population based metaheuristic) and GRASP algorithms are in the other extreme (single constructive metaheuristic). HS metaheuristics can be considered in the middle of these two alternatives, because they start with a population of solutions but usually end with a single solution. Some main details of the metaheuristics implementation are the following:

1. HS Algorithms: we ran 100 independent iterations of HS Algorithms. The number of groups and autonomous iterations were randomly selected, with a uniform distribution, in the intervals [$|V|/80$, $|V|/2$] and [1, 11] respectively.
2. GRASP: we ran 100 independents iterations of GRASP. The cut-off value α used to construct the RCL is selected randomly, with a uniform distribution in [0,1].

3. Memetic Algorithms: we ran 100 independent iterations. Each iteration had a population of 8 individuals, the maximum number of generations was 20, the probability of crossover was 0.8 and finally the probability of mutation was $1/|V|$.

All the metaheuristic were tested on the graphs Gxx shown in Table 1. These test problems were generated by Helmberg and Rendl using the graph generator described in [6]. They are *planar*, *toroidal* and *randomly* generated graphs of varying sparsity and size. The last two graphs types are in general non-planar. In these experiments, the graph sizes vary from 800 nodes to 3000 and their density from 0.17% to 6.12%.

The first three columns of Table 1 show respectively the *name* of the graph, the number of nodes n and the number of arcs m. The following three columns are the experiments for the three tested metaheuristic (Memetic algorithms, GRASP and HS algorithms). These columns are divided into three sub-columns: the maximum Max-Cut value found in the 100 independent iterations *cut*, the time spent in running these 100 independent iterations $T(s)$, and the *mean* and the standard deviation *sd*, respectively in these 100 independent iterations, shown in the third sub-column.

Right column shows the SDP value [6] that can be considered as an upper bound.

The experimental comparison of these metaheuristics shows a superior behavior of HS algorithms in relation with the computational complexity and quality of the approximate solutions obtained. Moreover, the quality of the obtained solutions using the memetic approach is also remarkable.

Table 1. Results for Helmberg´s instances [8]: solutions found with 100 independent iterations for Memetic, GRASP and HS metaheuristics. In bold is the maximum value for each instance.

Problem			Memetic Algorith			GRASP			HS Algorithm			
Name	n	m	Cut	T(s)	Mean (sd)	Cut	T(s)	Mean (sd)	Cut	T(s)	Mean (sd)	SDP
G1	800	19176	11546	1356	11466.4 (34.0)	11475	248	11369.8(39.6)	**11549**	221	11444.8(33.7)	12078
G2			**11546**	1340	11464.9(39.9)	11506	246	11376.2(37.2)	11501	215	11446.2(29.4)	12084
G3			11545	1336	11461.6(36.6)	11467	317	11374.3(34.2)	**11550**	226	11442.8(35.1)	12077
G11	800	1600	380	770	353.2(13.1)	506	371	482.0(10.0)	**546**	256	532.5(6.78)	627
G12			380	285	340.2(15.5)	500	362	481.2(8.01)	**540**	241	525.6(6.82)	621
G13			398	284	364.0(15.6)	530	335	500.0(9.38)	**568**	268	551.7(6.33)	645
G14	800	4694	2990	768	2961.9(14.1)	2983	418	2962.9(9.13)	**3014**	205	2986.6(9.05)	3187
G15			2970	760	2941.4(13.1)	2964	412	2941.6(10.5)	**2993**	208	2969.4(9.38)	3169
G16			2971	762	2947.5(12.2)	2967	411	2946.9(10.4)	**2996**	199	2973.5(8.63)	3172
G22	2000	19990	**13089**	5709	12987.7(55.5)	12971	7578	12884.4(33.6)	13061	1975	12993.7(34.9)	14123
G23			**13110**	7232	12988.9(54.9)	12998	7598	12887.5(40.5)	13098	2035	12997.0(36.9)	14129
G24			**13125**	7295	12996.7(51.9)	12978	7606	12890.1(39.2)	13089	1996	12990.3(37.8)	14131
G32	2000	4000	916	3262	846.0(26.3)	1220	4777	1193.6(12.7)	**1360**	3275	1330.9(10.9)	1560
G33			900	3223	835.1(31.5)	1212	4761	1179.8(12.5)	**1324**	3289	1299.9(13.0)	1537
G34			888	3266	812.3(28.8)	1228	4765	1188.5(17.5)	**1334**	3269	1308.8(12.1)	1541
G35	2000	11778	7478	3819	7408.1(31.5)	7456	6270	7422.0(14.1)	**7548**	2350	7487.7(18.2)	8000
G36			7450	3965	7399.0(28.0)	7445	6290	7417.7(14.0)	**7524**	2389	7482.2(18.2)	7996
G37			7465	3947	7411.8(30.9)	7456	6295	7426.7(13.7)	**7548**	2415	7488.7(19.7)	8009
G43	1000	9990	6553	1239	6494.3(28.7)	6500	981	6438.0(27.1)	**6561**	655	6490.9(22.8)	7027
G44			6532	1220	6490.0(24.5)	6497	922	6432.7(24.5)	**6540**	721	6488.5(22.9)	7022
G45			6545	1200	6485.4(31.0)	6499	952	6435.7(26.6)	**6564**	683	6484.9(24.4)	7020
G48	3000	6000	5008	7007	4869.6(84.6)	5996	12653	5991.2(13.3)	**6000**	7420	5931.2(74.5)	6000
G49			5024	6906	4859.5(81.0)	5996	12521	5988.3(22.0)	**6000**	7525	5930.7(65.6)	6000
G50			4988	9276	4848.3(81.3)	**5880**	12607	5877.2(7.40)	5880	7389	5840.8(41.6)	5988

7 Conclusions

This paper has introduced a hierarchical social algorithm to efficiently solve the Max-Cut problem. We have converted the search process into a social evolution one. This social paradigm exploits the power of competition and cooperation among different groups to explore the solution space. We have experimentally shown that the proposed HS algorithm significantly reduces the number of iterations of the search procedure. The memetic approach described has also given high quality solutions.

We propose as a future work a deeper study about the correlation between the performance of the algorithms and the main characteristics of the graph (type, density, etc.).

References

1. Burer, S., Monteiro, R.D.C., Zhang X.: Rank-two Relaxation heuristic for the Max-Cut and other Binary Quadratic Programs. SIAM Journal of Optimization, 12:503-521, 2001.
2. Dolezal O., Hofmeister, T., Lefmann, H: A comparison of approximation algorithms for the MAXCUT-problem. Reihe CI 57/99, SFB 531, Universität Dortmund, 1999.
3. Fernández F., Duarte A., Sánchez A.: A Software Pipelining Method Based on Hierarchical Social Algorithms, Proceedings of MISTA Conference, Vol. 1, pp 382-385, Nottingham, UK, 2003.
4. Fernández F., Duarte A., Sánchez A.: Hierarchical Social Algorithms: A New Metaheuristic for Solving Discrete Bilevel Optimization Problems, Technical Report ESCET/URF -DTF/ UPM. HSA1 Universidad Politécnica de Madrid, 2003.
5. Festa P., P.M. Pardalos, M.G.C. Resende, and C.C. Ribeiro: Randomized heuristics for the MAX-CUT problem, Optimization Methods and Software, vol. 7, pp. 1033-1058, 2002.
6. Goemans, M. X., Williams, D.P.: Improved Approximation Algorithms for Max-Cut and Satisfiability Problems Using Semidefinite Programming. Journal of the ACM.42:1115-1142 , 1995.
7. Goldberg. D.: Genetic Algorithms in Search, Optimization and Machine Learning. Addison-Wesley, 1989.
8. Helmberg, C., Rendl, F.: A Spectral Bundle Method for Semidefinite Programming. SIAM Journal of Computing, 10:673:696, 2000.
9. Karp, R.M.: Reducibility among Combinatorial Problems. In R. Miller J. Thatcher, editors, Complexity of Computers Computation, Prenum Press, New York, USA (1972).
10. Hadlock F. O: Finding a Maximum Cut of a Planar Graph in Polynomial Time. SIAM Journal on Computing 4 (1975) 221-225.
11. Lovàsz, L.: On the Shannon Capacity of a Graph, IEEE Trans. of Information Theory, IT-25:1-7, 1978.
12. Michalewicz, Z.: Genetic Algorithms + Data Structures = Evolution Programs. 3rd edn. Springer-Verlag, Berlin Heidelberg New York, 1996.
13. Michalewicz, Z., Fogel. D. B.: How to Solve it: Modern Heuristics, Springer-Verlag, Berlin, 2000.
14. Moscato P., Cotta C.: A Gentle Introduction to Memetic Algorithms. In Handbook of Metaheuristic. F. Glover and G. A. Kochenberger, editors, Kluwer, Norwell, Massachusetts, USA, 2003.

15. Resende M.G., Ribeiro C.: Greedy Adaptive Search Procedures. In Handbook of Meta-heuristic. F. Glover and G. A. Kochenberger, editors, Kluwer, Massachusetts, USA , 2003.
16. Resende M.G.: GRASP With Path Re-linking and VNS for MAXCUT, In Proceedings of 4th MIC, Porto, July 2001.
17. Spears, W. M.: Evolutionary Algorithms. The Role of Mutation and Recombination, Springer-Verlag, Berlin Heidelberg New York, 1998.
18. Shor, N. Z.: Quadratic Optimization Problems, Soviet Journal of Computing and System Science, 25:1-11, 1987.

On the Structure of Sequential Search: Beyond "No Free Lunch"

Thomas English

The Tom English Project
5401 50th Street I-1
Lubbock, Texas 79414 USA
Tom.English@ieee.org
http://www.TomEnglishProject.com

Abstract. Novel results are obtained by investigating the search algorithms predicated in "no free lunch" (NFL) theorems rather than NFL itself. Finite functions are represented as strings of values. The set of functions is partitioned into maximal blocks of functions that are permutations of one another. A search algorithm effectively maps each test function to a permutation of the function. This mapping is partitioned into one-to-one correspondences on blocks. A distribution on the functions is referred to as block uniform if each function has the same probability as all its permutations. For any search algorithm, the distributions of test functions and permuted test functions have identical Kullback-Leibler distance to the nearest block uniform distribution. That is, divergence from block uniformity is conserved by search. There is NFL in the special case of no divergence, i.e., the distributions are block uniform.

1 Introduction

Since 1995, "no free lunch" (NFL) in search of finite functions has been formulated in terms of algorithms that are sequential, deterministic, and non-redundant in the sense of never revisiting a domain point [1-2]. The present work is apparently the first to explore generally the mathematical structure (relationships between permutations, partitions, functions, algorithms, and probability distributions) of such search. The structure proves to be much less arcane than the confusion and debate over NFL would suggest. In fact, a single figure captures many of the salient features, including the necessary and sufficient condition for no free lunch (a correction to what has been published [3]). NFL is revealed to be essentially a minor manifestation of deeper attributes that ultimately should be of greater theoretical and practical significance.

Note that combinatorial optimization is typically treated as a search process, and proofs of NFL in optimization actually have established the far more general result of NFL in search [1-2]. The present work applies to optimization no less than past work has, but refers to search to indicate accurately the full generality of the results.

J. Gottlieb and G.R. Raidl (Eds.): EvoCOP 2004, LNCS 3004, pp. 95–103, 2004.
© Springer-Verlag Berlin Heidelberg 2004

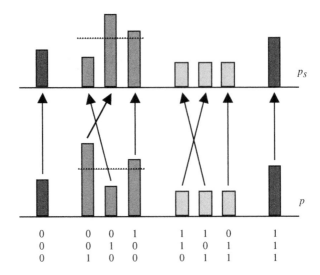

Fig. 1. Search algorithm s transforms the distribution of test functions, p, into the distribution of search results, p_s. Functions from $\{1, 2, 3\}$ to $\{0, 1\}$ are described as 8 strings of values grouped into 4 blocks. Algorithm s induces a one-to-one correspondence that associates each function with a permutation of itself (a function in its own block). The arrow pointing from the bar for 110 in the p graph to that for 101 in the p_s graph indicates that s observes values in the order 101 when the test function is 110. No arrow crosses from one block to another, and no two arrows point to the same bar. Thus within each block the p_s-probabilities are a permutation of the p-probabilities. If probabilities in the second block are set to the mean indicated by the dotted lines, $p_s = p$ for all search algorithms s.

Finite *test functions* are represented as strings of values. In Fig. 1, functions are from domain $\{1, 2, 3\}$ to codomain $\{0, 1\}$, and the function f with $f(1) = 1$, $f(2) = 0$, and $f(3) = 1$ is described 101. Search of a test function is formulated as a process of incrementally constructing a permutation of the domain. The permutation is initially empty, and domain points are added to the end of the permutation one-by-one. The choice of point is determined by the sequence of values associated with points already selected. The *search result* is obtained by applying the constructed permutation to the test function. That is, a search algorithm maps test function f to search result $j(f)$, where $j = d(f)$ is the constructed permutation and d is the sequential decision process. Fig. 1 gives a simple example in which $d(f) = 1, 3, 2$ for all functions f. Applying permutation $d(001)$ to 001 gives 010, and thus there is an arrow from the bar for 001 to that for 010.

Fig. 1 also illustrates a partition of the set of all test functions into maximal blocks of functions that are permutations of one another. Test functions and search results are formally indistinguishable, and the search result is always in the same block as the test function. Moreover, searches of distinct functions yield distinct results. Thus the mapping from test functions to search results can be partitioned into one-to-one correspondences on the blocks. The partition of the set of test functions and the related partition of the mapping permit simple reasoning about search in terms of blocks. This is illustrated in formal derivations in Section 3.

If all test functions within a block have equal probability, all search results for that block must have the same probability, irrespective of the search algorithm. This holds for three of the four blocks in Fig. 1. But if two functions within a block do not have the same probability, there exists a permutation (a simple search algorithm) that maps one to the other, and the distribution of search results differs from the distribution of test functions. A distribution that is uniform within each block is called *block uniform*. The distributions of Fig. 1 can be made block uniform by setting all probabilities in the second block to the mean indicated by the dotted lines. The construction gives the same result for both distributions. For typical measures of information distance, the depicted distributions are closer to the constructed distribution than to any other block uniform distribution, and both are the same distance from the constructed distribution.

From the foregoing, every search algorithm transforms a block uniform distribution of test functions into an identical distribution of search results. Also, any distribution of test functions that diverges from block uniformity is transformed into a different distribution of search results by some search algorithm. But the two distributions are identical if the search algorithm is the identity permutation. It follows that search results are identically distributed for all search algorithms if and only if the distribution of test functions is block uniform. That is, a block uniform distribution of test functions is necessary and sufficient for "no free lunch" in search.

The structural attributes of search introduced here, and some others as well, are demonstrated rigorously below. The following sections present formal preliminaries, derivations of main results, discussion, and conclusions.

2 Preliminaries

Test functions from $X = \{1, ..., N\}$ to $Y = \{1, ..., M\}$ are represented as strings y in $Y = Y^N$, with y_n the value at point $n \in X$. Let p and q be probability distributions on Y. The Kullback-Leibler distance of p from q is

$$D(p\|q) = \sum_{y \in Y} p(y) \log \frac{p(y)}{q(y)}. \tag{1}$$

$D(p \| q) \geq 0$, with equality if and only if $p = q$ [4].

Let J be the set of all permutations $j = j_1, ..., j_N$ of X. Each $j \in J$ is interpreted as an invertible mapping on Y, $j(y) = y_{j_1} \cdots y_{j_N}$, with $\iota = 1, ..., N$ the identity function. For all $y \in Y$, let *block* $[y] = \{j(y): j \in J\}$, and let $\pi = \{[y]: y \in Y\}$.

A *search algorithm* $s: Y \to Y$ with *decision process* $d: Y \to J$ is defined $s(y) = d(y) \circ \iota(y)$, with $d_n(y)$ determined by $s_1(y), ..., s_{n-1}(y)$, $n = 1, ..., N$. It follows immediately that $s(y) \in [y]$, and that if d is constant, $s \in J$. Let S be the set of all search algorithms, with $J \subseteq S$.

3 Main Results

The set of all permutations of function y has been named a *block*, and π is by definition the collection of all blocks of functions, but it remains to be established that π is indeed a partition of the set of functions. The following proof shows that every function is in exactly one block, and that no block is empty.

Theorem 1. π is a partition of Y.

Proof. For every $y \in Y$, block $[y] \in \pi$ contains $y = \iota(y)$. Thus the blocks are nonempty and exhaustive. For all $w, y \in Y$, $y \in [w]$ implies $y = j(w)$ and $w = j^{-1}(y)$, with $w \in [y] = [w]$. Thus the blocks in π are disjoint. It follows that π is a partition.   □

The next theorem establishes that a search algorithm, which has been defined as a mapping for convenience, brings test functions into one-to-one correspondence with search results, and respects partition π in the sense that each test function is associated with a search result in its own block. Thus the search algorithm itself can be partitioned into one-to-one correspondences, one for each block of π.

Theorem 2. Let $s \in S$. Then s is a one-to-one correspondence with $s([y]) = [y]$ for all $y \in Y$.

Proof. Let $y \in Y$, and let d be the decision process of s. For all w in $[y]$, $s(w) \in [w] = [y]$, so $s([y]) \subseteq [y]$. It follows that $s([y]) = [y]$ and $s(Y) = Y$ if s is one-to-one. Suppose $s(w) = s(y)$ and $w \neq y$ for some $w, y \in Y$. Then $d_n(w) \neq d_n(x)$ for some $n \in X$. But $d_n(w)$ and $d_n(x)$ are determined by $s_1(w) \ldots s_{n-1}(w) = s_1(y) \ldots s_{n-1}(y)$, a contradiction.   □

Corollary. $\pi_s = \{s \cap [y]^2 : y \in Y\}$ is a partition of s into one-to-one correspondences on the blocks of π.

Proof. It follows from $s([y]) = [y]$ that for all $y \in Y$ both $(y, s(y))$ and $(s^{-1}(y), y)$ are in $s \cap [y]^2$. Thus the blocks of π_s are nonempty and exhaustive, with each $s \cap [y]^2$ an invertible mapping on $[y] \in \pi$. Disjoint $[y] \in \pi$ implies disjoint $s \cap [y]^2$, and π_s is a partition of s.   □

In the following, the block uniform distribution "nearest" to a given distribution of functions is defined. The "nearest" property is neither required nor proved in the present work, but it should be intuitive that setting the probabilities within each block of the given distribution to the mean for the block constitutes minimal adjustment.

Definition. Probability distribution p on Y is *block uniform* if and only if $p = \bar{p}$, where

$$\bar{p}(y) \equiv \bar{p}([y]) = \frac{1}{|[y]|} \sum_{w \in [y]} p(w) \tag{2}$$

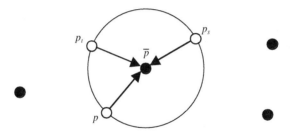

Fig. 2. The distribution of test functions p is on a non-Euclidean hypersphere with its corresponding block uniform distribution $\overline{p}$ at the center. The distributions of search results for all search algorithms s and t lie on the hypersphere, and $\overline{p} = \overline{p}_s = \overline{p}_t$.

for all $y \in Y$. Consistency follows from Theorem 1. It follows easily that p is block uniform if and only if $p(w) = p(y)$ whenever $w \in [y]$.

As indicated in Fig. 1, the probability distribution of search results is a transformation of the distribution of test functions. The search algorithm induces the transformation. The relationship between the two distributions is simple. The distribution of search results is the composition of the distribution of test functions with the inverse of the search algorithm. (Recall that a one-to-one correspondence is an invertible function.)

Definition. For $s \in S$, let $p_s = p \circ s^{-1}$ be a probability distribution on Y. Consistency follows from the fact that s is a one-to-one correspondence on the sample space of p (Theorem 2).

Next it is established that the block uniform distribution corresponding to the distribution of search results is that corresponding to the distribution of test functions. The proof hinges upon the fact that the mean probability of test functions within a block is the mean probability for search results within the block.

Theorem 3. Let $s \in S$. Then $\overline{p} = \overline{p} \circ s^{-1} = \overline{p}_s$.

Proof. For all $y \in Y$, $\overline{p}([y]) = \overline{p}(s^{-1}([y]))$ by Theorem 2, and thus $\overline{p} = \overline{p} \circ s^{-1}$. Now $\overline{p}_s = \overline{p} \circ s^{-1}$ because

$$\frac{1}{\left\| [y] \right\|} \sum_{w \in [y]} p \circ s^{-1}(w) = \frac{1}{\left| s^{-1}([y]) \right|} \sum_{w \in s^{-1}([y])} p(w) \qquad (3)$$

for all $y \in Y$, by Theorem 2. □

The following establishes that the distributions of test functions and search results are equidistant from their corresponding block uniform distribution (see Fig. 2). The proof depends on the identities $s(Y) = Y$ and $s^{-1}(s(y)) = y$, which hold for all search algorithms s.

Theorem 4. Let $s \in S$. Then $D(p_s \| \overline{p}_s) = D(p_s \| \overline{p}) = D(p \| \overline{p})$.

Proof. The first equality follows from $\bar{p} = \bar{p}_s$ (Theorem 3). For all $y \in Y$, $p(y) = p_s(s(y))$, and

$$D(p \| \bar{p}) = \sum_{y \in Y} p_s(s(y)) \log \frac{p_s(s(y))}{\bar{p}(y)}$$

$$= \sum_{y \in s(Y)} p_s(y) \log \frac{p_s(y)}{\bar{p}(s^{-1}(y))}$$

$$= \sum_{y \in Y} p_s(y) \log \frac{p_s(y)}{\bar{p}(y)}$$

$$= D(p_s \| \bar{p}) \tag{4}$$

because $s(Y) = Y$ (Theorem 2) and $\bar{p}(s^{-1}(y)) = \bar{p}_s(y)$ (Theorem 3).   □

The following theorem and corollaries effectively state that a block uniform distribution is necessary and sufficient for no free lunch in search. If all search algorithms, including permutations, induce a distribution of search results identical to the distribution p of test functions, then $p(y) = p_j(y)$ and $p(y) = p(j^{-1}(y))$ for all permutations j and test functions y. This implies that all test functions that are permutations of one another have the same probability, and the distribution of test functions is block uniform. Conversely, if the distance of the distribution of test functions from the nearest block uniform distribution is zero, then the distance of the distribution of search results from the distribution of test functions must be zero for all search algorithms.

Theorem 5. It holds that $p_s = p$ for all $s \in S$ if and only if $p = \bar{p}$.

Proof. If $p_s = p$ for all $s \in S$ then $p_j(y) = p(y)$ and $p(y) = p(j^{-1}(y))$ for all $j \in J \subseteq S$ and $y \in Y$. It follows that $p(w) = p(y)$ for all $w \in [y]$, and $p = \bar{p}$. Conversely, if $p = \bar{p}$ then $D(p \| \bar{p}) = 0$. By Theorem 4, this implies that $D(p_s \| \bar{p}) = 0$ for all $s \in S$, and thus $p_s = p = \bar{p}$ for all $s \in S$.   □

Corollary. It holds that $p_s = p_t$ for all $s, t \in S$ if and only if $p = \bar{p}$.

Proof. Note that $p = p_\iota$ for identity permutation $\iota \in J \subseteq S$. The result follows by symmetry and transitivity of the equality relation on probability distribution functions.   □

Corollary. $D(p_s \| p) = 0$ for all $s \in S$ if and only if $D(p \| \bar{p}) = 0$.

The last theorem illustrates use of the block structure of search to obtain a simple answer to a difficult question. Shortly after NFL theorems [1] became available on the Internet, Park [5] interpreted NFL in light of the theory of Kolmogorov complexity, pointing out that almost all functions are algorithmically random. Hartley [6] later made the similar observation that most functions are so complex (require so many bits to describe) that they cannot be realized physically, and raised the question of whether a uniform distribution on the set of all functions with short descriptions offers no free lunch. The description of function y is, loosely, the shortest program that outputs y and halts. The length in bits of the program is $K(y)$, the *Kolmogorov complexity* of y. (See [7] for details.) Under the simplifying assumption that the domain and codomain of functions are identical, it is shown that there is a block that contains a function of low complexity and many functions of high complexity. If functions of high complexity have probability zero, and the function of low complexity does not, then the distribution within the block is not uniform.

Theorem 6. Assume large $M = N$, and let probability distribution p be uniform on $\{y \in Y: K(y) \leq \theta\}$. Then p is not block uniform if $\log^* N + c \leq \theta \leq \log N!$, where c is a small constant.

Proof. Let $y \in Y$, with $y_n = n$, $n = 1, \ldots, N$. The value of N can be encoded in about $\log^* N$ bits [7], and the length of the shortest code to write y, given N, is a small constant c. Thus $K(y) = \log^* N + c$. But the number of permutations of y is $N!$, so the mean value of $K(w)$, $w \in [y]$, is at least $\log N!$ For sufficiently large N, there is some $w \in [y]$ such that $\log^* N + c \leq \log N! < K(w)$. Then p is not block uniform for $\log^* N + c \leq \theta \leq \log N!$, because $p(y) > p(w) = 0 \neq \overline{p}(w)$. □

4 Discussion

Corresponding to the partition of the set of test functions into blocks of functions that are identical under permutation is a partition of the search algorithm into one-to-one correspondences on the blocks. The behavior of a search algorithm can be analyzed on a block-by-block basis. This is not to say, however, that the search algorithm behaves independently for different blocks. For instance, the choice of the first point to enter into the permutation is constant.

The relationships between distributions of test functions and search results can be stated in geometric terms. (The space is non-Euclidean because the Kullback-Leibler distance is not symmetric and does not respect the triangle inequality). For test-function distribution p there is a hypersphere containing all distributions that are the same distance from $\overline{p}$ as p. The search-result distribution p_s induced by search algorithm s is a point on that hypersphere. Both p and p_s are closer to $\overline{p}$ than any other block uniform distribution. All search algorithms yield the same distribution of search results if and only if the radius of the hypersphere is zero.

In unpublished work, Neil and Woodward [9] have independently identified the block uniform distribution and proved its necessity and sufficiency for NFL. Other investigators, in published work [3], have claimed that a uniform distribution on a set of test functions closed under permutation is necessary and sufficient. This is equivalent to a block uniform distribution with all nonzero probabilities identical. The condition is sufficient, but not necessary. The number of distributions satisfying the condition is finite, and it was established long ago that an uncountable set of distributions offers no free lunch [8]. The block uniform distributions are indeed uncountable. To construct one, first apportion the probability mass freely among the blocks. Then divide the probability mass of each block equally among its members. The number of ways to do the first step is uncountable, and the way to do the second step, given the first, is unique.

Theorem 6, while it does illustrate simple reasoning in terms of the block structure of search, is not entirely in keeping with the thrust of the paper. It suggests a crisp dichotomy of distributions that are block uniform and those that are not, when Theorem 4 suggests that dichotomous characterization is neither necessary nor generally appropriate. Future work should quantify (perhaps approximately) the divergence from block uniformity of a distribution uniform on the set of all low-complexity functions. This would complement "almost no free lunch" results derived in [10].

Extension of the results developed here to randomized search is straightforward, but rigorous treatment is left for future work. A randomized search algorithm essentially makes a random selection of deterministic search algorithm, and then executes it. One may define the distribution of deterministic search algorithms in such a way that the randomized algorithm performs a random walk of the function domain. In this case the search results follow the block uniform distribution $\bar{p}$. More generally, randomized search places the distribution of search results on or inside the hypersphere of Fig. 2.

An open problem is, given test function distribution p, how to determine the maximum Kullback-Leibler distance between distributions p_s and p_t of search results for any search algorithms s and t. It would be useful to express that distance as a function of $D(p \parallel \bar{p})$. Future work should also explore the implications of isomorphism in search algorithms. Search algorithms s and t are *isomorphic* if $t \circ j = s$ for some permutation j. That is, the mappings are identical up to a renaming of domain points. There is clearly a relationship to the block structure of search, but the implications remain to be explored. Finally, the results of the present work should be applied to optimization in future studies.

It is likely that the salient structure of sequential search is present in some apparently dissimilar forms of search. For instance, the typical implementation of an evolutionary (parallel, stochastic) algorithm is a deterministic, sequential program, and one may abstract from the program a sequential algorithm. Addition of steps to prevent selection of a point more than once yields a sequential search algorithm of the type studied here. Whether or not this approximation conceals relevant aspects of the behavior of the evolutionary algorithm depends upon the goals in analysis and the incidence of revisited points. But given the relative ease in understanding sequential search, the approximation merits consideration.

5   Conclusion

The present work indicates the value of studying sequential, deterministic, non-redundant search as a topic unto itself, rather than focusing upon the phenomenon of "no free lunch." It has been demonstrated that there is a prominent block structure in search that simplifies analysis. Consideration of this block structure leads to the identification of block uniform distributions as distinguished points in the space of probability distributions on functions. Any distribution of test functions lies on a non-Euclidean hypersphere with the nearest block uniform distribution at the center. A search algorithm induces a distribution of search results on the same hypersphere, and that distribution is closer to the center than to any other block uniform distribution. There is no free lunch in the special case that the radius is zero.

These results carry analysis of search and combinatorial optimization into new territory. They apply to all distributions, and support quantitative characterization. They show that NFL, while interesting, is a modest part of the overall picture. The block structure of search gives rise to NFL, but is in no sense a generalization.

References

1. Wolpert, *D*., Macready, W.: No Free Lunch Theorems for Search. Santa Fe Institute, SFI-TR-02-010 (1995)
2. Wolpert, D., Macready, W.: No Free Lunch Theorems for Optimization. *IEEE Trans. Evolutionary Computation* 1 (1997) 67-82
3. Schumacher, C., Vose, M. D., Whitley, L. D.: The No Free Lunch and Problem Description Length. In: Spector, L., Goodman, E. D., Wu, A., Langdon, W., Voight, H. M., Gen, M., Sen, S., Dorigo, M., Pezeshk, S. Garzon, M. H., Burke, E. (eds.): *Proc. of the Genetic and Evolutionary Computation Conference* (GECCO-2001), Morgan Kaufmann, San Francisco (2001) 565-570
4. Cover, T. M., Thomas, J. A.: *Elements of Information Theory*. Wiley & Sons, New York (1991)
5. Park, K. In: Genetic Algorithms Digest 9, http://www.aic.nrl.navy.mil/galist/digests/ v9n10 (1995)
6. Hartley, R. In: Genetic Algorithms Digest 15, http://www.aic.nrl.navy.mil/galist/digests/v15n17 (2001)
7. Li, M., P. Vitányi, P.: *An Introduction to Kolmogorov Complexity and Its Applications*. Springer Verlag, New York (1997)
8. English, T. M.: Evaluation of Evolutionary and Genetic Optimizers: No Free Lunch. In: Fogel, L. J., Angeline, P. J., Bäck, T. (eds.): *Evolutionary Programming V: Proc. 5th Ann. Conf on Evolutionary Programming*, MIT Press, Cambridge, Mass. (1996) 163-169
9. Neil, J., Woodward, J.: The Universal Distribution and a Free Lunch for Program Induction. Unpublished manuscript.
10. Droste, S., Jansen, T., Wegener, I.: Optimization with Randomized Search Heuristics: The (A)NFL Theorem, Realistic Scenarios, and Difficult Functions. *Theoretical Computer Science* 287 (2002) 131-144

Dealing with Solution Diversity in an EA for Multiple Objective Decision Support – A Case Study

Alvaro Gomes[1,2], Carlos Henggeler Antunes[1,2], and Antonio Gomes Martins[1,3]

Department of Electrical Engineering and Computers
University of Coimbra, Polo II
3030-290 Coimbra, Portugal
{agomes,ch,amartins}@deec.uc.pt
INESC Coimbra, R. Antero de Quental 199,
3000-033 Coimbra, Portugal

Abstract. The characteristics of the search space (its size and shape as well as solution density) are key issues in the application of evolutionary algorithms to real-world problems. Often some regions are crowded and other regions are almost empty. Therefore, some techniques must be used to avoid solutions too close (which the decision maker is indifferent to) and to allow all the regions of interest to be adequately represented in the population. In this paper the concept of δ-non-dominance is introduced which is based on indifference thresholds. Experiments dealing with the use of this technique in the framework of an evolutionary approach are reported to provide decision support in the identification and selection of electric load control strategies.

1 Introduction

Most real-world problems to be tackled by evolutionary algorithms (EA) are generally very complex. This complexity stems, for instance, from the combinatorial nature of those problems namely whenever multiple, incommensurate and conflicting objectives are at stake. In these cases, in general, there is no knowledge about the shape and the morphology of the search space. The situation becomes even more complex when the search space is very large and there is no information that allows its size to be reduced. In the presence of irregular search spaces and variable solution distribution density, finding good compromise (non-dominated) solutions can be a hard task. In general, some regions are crowded and other regions are almost empty, asking for some filtering technique to be used in order to avoid the presence in the population of solutions that are too close to other solutions and to allow all the regions of interest for the decision maker (DM) to be adequately represented in the population. In this context, an important issue is what approach to use for guaranteeing the diversity of the population and avoiding very close neighbours.

Usually, there are three main criteria to evaluate the performance of an EA: convergence to the Pareto Optimal Front (POF), diversity and good distribution of the population along the POF for adequately representing it (Zitzler et al., 2000). However, in multiobjective (MO) models it is not straightforward to implement a

J. Gottlieb and G.R. Raidl (Eds.): EvoCOP 2004, LNCS 3004, pp. 104–113, 2004.

metric able to deal with all these criteria simultaneously. In fact, these three criteria also present some conflicting aspects, and several approaches have been proposed to deal with each of them. For example, in order to assure convergence to the FOP an elitist behaviour is generally used allowing for the maintenance of the fittest individuals in the population (Zitzler and Thiele, 1998; Knowles and Corne, 2000; Deb et al., 2000). Fitness sharing, mating constraints and crowding are among the more "traditional" tools used to maintain the diversity in the population. The main idea underlying fitness sharing is that individuals of the population in the same region of the search space are competing for the finite resources existing in the region, and thus more individuals means less resources for each one (Goldberg, 1989). In genetic populations the sharing of the resources among neighbours leads to some degradation of their individual fitness, that is the level of degradation of the fitness of an individual depends on the number of its neighbours. Restrictions to mating can be used both for diversity preservation and avoiding the generation of lethal (very low performance) individuals that can occur when very distinct individuals mate.

Other approaches related with the diversity (and elitism) issues have been suggested by different authors. DeJong (1975) proposed a partial replacing method in which the replaced individual is the most similar (that exists in a group of individuals randomly chosen from the population) with the new child, thus allowing some diversity to be maintained in the population. Fonseca and Flemming (1998) introduced a small number of random immigrants in the population. Zitzler (1998) uses an external population, in the framework of SPEA, in which some of the non-dominated solutions found until the current generation are stored. In order to avoid the size of this population to grow beyond a pre-defined size, S, a clustering technique is used in which the entire population is divided into S clusters and the centroid of each cluster is used as a representative. PAES (Knowles and Corne, 2000) also uses an external population of non-dominated solutions which accepts new members if there is any free place or the candidate dominates one or more solutions in the archive, and these dominated solutions are erased from the archive. When the candidate neither dominates nor is dominated by any individual in the archive (meaning that the candidate and the solutions in the archive belong to the same non-dominated front), the solution in the less crowded region is chosen. Here, the objective space is divided (grid) into regions and the candidate is accepted if it lives in a less crowded region (cell of the grid) than any other member of the archive, otherwise it is rejected.

Elitism is implemented in NSGAII through the evaluation of the population built up with the parents and offspring, the new generation consisting of the non-dominated individuals in this combined population. When there is a need to choose between two or more candidates the one in the less crowded region is selected. The use of a grid as a way to reduce the size of the non-dominated front is also mentioned in Coello et al. (2002). Osyczka (2002) makes use of a filtering process which allows the removal of solutions, from the Pareto set, which are closer to other solutions in any of the objectives. Laummans et al. (2002) define the ε-dominance concept used to identify an ε-approximate Pareto set which, while representative of the POF, has also a smaller number of solutions. The strategy used by the authors guarantees both the convergence towards the Pareto Optimal Front and a set of well distributed solutions. In each iteration, a new solution is generated and the archive, containing a subset of all the solutions generated so far, is updated. The new solution is accepted only if it is not ε-dominated by any other solution in the archive. If the solution is accepted, all the dominated solutions are erased from the archive. The ε-dominance concept makes use of a tolerance in each objective.

This paper addresses the problem of the existence of very close neighbours in the population eventually contributing to a lack of diversity and a bad spreading over the Pareto Front. This can be a key issue in real-world problems, in which the fitness function makes use of actual data provided by physical devices or by models simulating the behaviour of real systems. In the work reported in this paper the selection operator performs a vital role by identifying so-called δ-non-dominated solutions. A δ-non-dominated solution works as a representative of a set of non-dominated solutions existing in a region or cluster. However, instead of splitting up the search space in clusters, the distinction between solutions is done resorting to thresholds. The DM is asked to specify the indifference thresholds δ_j in each evaluation dimension j (objective function) beyond which two solutions are considered distinct for that dimension, that is $\left| g_j(x) - g_j(y) \right| \leq \delta_j$ where $g_j(x)$ and $g_j(y)$ are the performances of solution x and y with respect to objective j (figure 1).

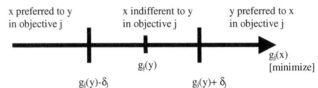

Fig. 1. Indifference thresholds for evaluation of solutions x and y in objective j.

These indifference thresholds allow the identification of a hyper-volume around each non-dominated solution. All the solutions within the hyper-volume of other solution are considered equivalent from the DM's point of view, and therefore it is not necessary to keep all of them in the population for the purpose of providing the DM with meaningful information on which a final decision could be based. Instead of splitting up the search space into regions or cells of a grid, the δ-non-dominance concept takes the indifference thresholds into account to avoid the presence of indistinguishable individuals in the set of non-dominated solutions.

2   The Selection of Load Management Actions: A Multiobjective Combinatorial Problem

Load control strategies, in the framework of Load Management (LM) approaches in power systems, have been used by several electric utilities because of their potential attractiveness. In these approaches some end-use loads are switched off and on during a given period of time (usually several hours), thus changing the level and the shape of demand. Besides the traditional benefits, such as cost reduction, reduction of air pollution, decrease of overseas fuel dependence, and increase in power system reliability, the increase in revenues is the main attractiveness of LM in the ongoing restructuring and liberalisation trends in the energy sector. The scenarios of total unbundling of integrated utilities, where several entities (generators, retailers, costumers) appear in each branch of activity, seem even more attractive for the dissemination of energy services activities, such as LM. As a consequence of the unbundling of the sector and the appearance of several entities, the impacts of such actions shall be evaluated and assessed at different demand aggregation levels. For instance, a retailer can be interested in the evaluation of the local effects on demand of load control actions implemented over some loads of a group of commercial consumers and, at the same time, he/she can be interested in the assessment of the impact on the total revenues or the global demand (of all the customers of this retailer). Therefore, a tool to provide decision support in such activities must be flexible and accurate to adequately evaluate the impacts of LM actions at different levels of demand aggregation.

An entity pursuing LM activities, the main business of which is selling electricity (either a distribution company or a retailer / marketeer dealing only with financial flows), faces several and conflicting aspects that must be explicitly considered and assessed in order to increase the success in the implementation of such actions. Some of the objectives that are pursuit are: minimization of maximum demand (at different levels of aggregation), minimization of eventual discomfort caused to customers, maximization of revenues, and minimization of power losses (Gomes et al., forthcoming). Load control actions (LCA) are usually applied over sets of groups of loads of the same type, during some time. In this interval of time, loads are turned *off* and *on* during some periods. While the loads belonging to a given group are controlled simultaneously, the *on/off* periods for all groups under control may not be coincident in order to avoid some undesirable effects. One of these is the payback effect, due to the restoration of power to loads at the same time after an interruption.

The problem of LCA identification means that each *off* period must be quantified (how long) and located in time (when). This should be done for all the groups under control. One control strategy encompasses all control actions applied over all the groups under control, and the identification of all *on/off* periods is a complex task. By allowing different durations of *off* and *on* periods there is an increase in flexibility in load control leading, in general, to better results. However, the problem complexity also increases. Generally, the identification (duration and location) of these periods is done empirically, based on field experiments or expensive and time consuming pilot programs. Some attempts to use linear and dynamic programming have also be done (Chen et al., 1995; Ng and Sheblè, 1998). Some of them, impose restrictions to the *on/off* periods or a very rigid schedule.

The decision variables are binary ones (each control strategy is applied or not to a group of loads in a time interval). The search space is of dimension $NS*NG$, where NG is the number of groups under control and NS is the number of solutions.

In this work the demand of controlled loads with and without LCA results from Monte Carlo simulations based on physically-based load models (PBLMs) that have been experimentally validated (Gomes et al., 1999). These models reproduce the behaviour of the simulated groups of loads with a minute time resolution, during one day. PBLMs are detailed models allowing to capture the effects of LCA on the demand of the controlled loads, thus making this simulation a very effective process with respect to the evaluation of changes in demand, even though it is a time consuming task. In the study reported in this paper there are 1860 loads under control, grouped in 20 groups. The objectives considered in the evaluation of load control strategies are: minimization of peak power demand at aggregate level (PA1), minimization of maximum demand at two different less aggregate level (PD1 and PD2), minimization of discomfort caused to the customers (maximum continuous time interval and total time comfort thresholds are violated), power losses minimization (L) and maximization of revenues (R). For details about the multiobjective mathematical model see Gomes et al. (forthcoming)

3   An Evolutionary Approach to Identify Load Control Actions

In EAs, evolution takes place based on appropriate genetic operators that select, recombine and mutate individuals and also on a adequate evaluation function. In this work the performance assessment of individuals is carried out through the PBLMs, which are able to simulate the demand of the loads under control when LCAs are applied to the loads. After a random initial generation is created by the EA, the iterative process proceeds as follows. The PBLMs compute the demand of all the loads according to the different on/off actions codified in each individual of the population. The obtained results (for each objective) are used by the EA to assess the merit of each individual. The fitness of each solution is based on the Pareto optimality concept, but allowing a tolerance value (pre-specified by the DM) in each objective when testing the non-dominance relations. First, the non-dominated solutions in the population are identified. If very close solutions exist then only the ones that are sufficiently distant, according to the tolerance values, are considered as δ-non-dominated solutions. These solutions are assigned the rank 1. The process is repeated until every solution is assigned a rank. Crossover, mutation and selection of the parents of the next generation follow. The process is repeated until some stop condition is reached. The stop condition can be a maximum number of generations or the quality of the individuals in the population is satisfactory according to the DM's judgement. Details about the selection operator are provided in the next section.

Since PBLMs simulate the demand of loads within a one minute time basis each control strategy is also implemented with a one minute resolution in order to increase the flexibility and thus improve the results of LM. The representation used in the EA for the LCA applied to each group is a binary string. Since one decision variable represents one control strategy then each individual in the population has the length *1440 * NG* (1440 minutes of the day times the number of groups). In order to prevent a high frequency of the *on/off* switching applied to loads, the minimum duration of both *on* and *off* periods is 5 minutes. Figure 2 shows the relationship of the genotype and phenotype spaces as well as the link with the solutions fitness assessment.

In our case study, as in most real-world applications, the shape and the morphology of the search space are unknown. Moreover, due to the number of objectives and the size of search space these are generally multi-modal problems.

Fig. 2. Genotype and phenotype spaces.

Each cycle of the iterative process encompasses one run of the EA and one run of the PBLMs. It takes about 7.9 seconds (Pentium III, 733MHz). PBLMs take about 99,2% of the time spent in each cycle. PBLMs used in the fitness assessment must run for every load present in each group and for every control strategy codified in each individual of the population. As in many real-life applications the link with the physical world is a time consuming process. There is a compromise between the need of an adequate number of solutions and the available time for obtaining the results. The size of population is 20. Thus, the search space is defined by 400 binary variables (20 groups of loads x 20 individuals in the population) and seven evaluation dimensions operationalized by objective functions (Gomes et al., forthcoming).

4   Identification of the δ-Non-dominated Set

Some approaches to the identification of a reduced Pareto set and different methods to deal with diversity issues have been already referred to in section I. This is an important issue in our problem, since the Pareto front is unknown and it is impossible to find all its points.

When different control actions are applied to the same group of loads the resulting power demand is also different (demand level and/or pattern). The number of different possibilities of load control actions over one group is very large leading to a multitude of different demand patterns. These figures greatly increase with the number of load groups. This means that individuals exist in the population that are too close to other individuals. This situation is even more plausible during the convergence process whenever some of the axes of the search space are real valued. This contributes for a low diversity, although some techniques can be used to avoid this situation, such as fitness sharing. This issue asks for some kind of filtering to avoid solutions the DM is indifferent to. For instance, the value for the objective function related with the maximum peak power demand at aggregate level before a LCA is applied is 31 632 kW. Results such as 30 590.5 kW or 30 590 kW frequently arise in the population, this type of differences being meaningless for the DM. Similar situations occur in the other evaluation dimensions. The DM is therefore offered very close solutions which are indistinguishable from a practical point of view. Fitness sharing can give a positive contribution to overcome this problem, as well as some of the approaches referred to in section I could be used. However, there is no a priori knowledge about the shape of the search space, and the distribution of compromise (non-dominated) solutions along the search space is unknown and expectedly irregular, leading to a situation in which some regions are crowded and some other regions are almost empty. Therefore, a different approach has been used in this work.

In order to deal with this issue a relaxed non-dominance definition is proposed. It consists in considering indifference thresholds in each dimension when the non-dominance definition is being tested, thus avoiding the existence in the population of very close solutions. This δ-non-dominance concept requires from the DM the specification of indifference thresholds in each evaluation dimension. These thresholds are more intuitive for the DM than to specify a distance in a multidimensional space. The idea is that if two or more solutions are inside the hyper-volume (defined resorting to the indifference thresholds in each dimension) of each other then only one of them should be identified as "non-dominated", thus acting as a representative of all the solutions in the hyper-volume. That is, if one solution, x', is inside the hyper-volume of the solution x'' defined by $x'' \pm \delta$ then only one solution should be part of the population. $\delta = (\delta_1, \delta_2, ..., \delta_n)$ is the vector of indifference thresholds (values in each dimension for which the DM is indifferent between two solutions).

Let $x' = (x'_1, x'_2, ..., x'_n)$ and $x'' = (x''_1, x''_2, ..., x''_n)$ be two non-dominated solutions, in the space of objective functions, and $\delta = (\delta_1, \delta_2, ..., \delta_n)$ the vector of indifference thresholds.

x' δ-dominates x'', $x' \succ_\delta x''$, iff:

1. $x''_i \in \left[x'_i - \delta_i, x'_i + \delta_i \right]$, $i = 1, ..., n$

2. x' is better than x'' for more objectives than the ones in which it is worse.

However, this relationship is not transitive. Therefore, some δ-non-dominated solutions may not be identified as such in a first screening of this type of solutions. In order to avoid this situation the δ-dominance concept must be iteratively verified.

Let $P^{(t)}$ be the current population in generation t. F_δ is the set of δ-dominated solutions and F_δ^* is the set of δ-non-dominated solutions. At the beginning F_δ and F_δ^* are empty sets.

- $\forall i : x^i \in P^{(t)}$
  - if $\exists x^j, j \neq i : x^j \succ_\delta x^i$ then $F_\delta^* := F_\delta^* + x^i$
  - else $F_\delta = F_\delta + x^i$
- Repeat
  - $\forall i : x^i \in F_\delta$
    - if $\exists x^j \in F_\delta^* : x^j \succ_\delta x^i$ then $F_\delta := F_\delta - x^i$
  - $\forall i : x^i \in F_\delta$
    - if $\exists x^j \in F_\delta : x^j \succ_\delta x^i$ then $F_\delta^* := F_\delta^* + x^i$
- until $F_\delta = 0$.

The extra cpu effort due to this calculation is insignificant when compared with the cpu time necessary for running the PBLMs, and it is used only in the comparison of two non-dominated solutions close to each other.

5   Case Study: Some Results

In this section the results of three implementations of the EA referred to in section III are analysed and compared. The main objective is to compare the efficacy of each alternative implementation in avoiding the presence in the population of individuals that are too close to others (despite they are non-dominated). This is an undesirable situation since it can result in reduced diversity in the population and it is difficult for the DM to make a distinction between solutions that are not different beyond given threshold levels.

The performance evaluation of the different implementations is carried out by using the *surface attainment* concept, which has been proposed in Fonseca and Fleming (1996) for comparing two sets of solutions. Knowles and Corne (2000) use a pair of values *(a, b)*, in which *a* represents the percentage of the space (lines) in which the first algorithm performs better that the second algorithm and *b* represents the percentage of the space (lines) in which the second algorithm performs better than the first one. If the set of lines cover the whole front then the *surface attainment* method allows us to deal with the three issues raised in the performance assessment: distance to Optimal Pareto Front, distribution and diversity of the solutions. Knowles and Corne (2000) further extended this analysis to more than two algorithms through the use of two "statistics", *unbeaten* and *beat all* by doing pairwise comparisons, as it is done for two algorithms, but now using all the points resulting from all the algorithms being compared for the "construction" of the Pareto front. *Unbeaten* is the percentage of the Pareto front where the algorithm is not beaten by any other algorithm, and *beats all* is the percentage of the front where the algorithm beats all other algorithms.

In this paper, the Mann-Whitney test at 95% confidence level and 13 runs for each algorithm have been used in the framework of the surface attainment method. The indifference thresholds used in the study are the following: maximum aggregate demand: 25 kW, maximum demand at both less aggregate level: 5 kW, total number of minutes in which the temperature is above/below the threshold level: 100 minutes, maximum interval in which the temperature is above/below the threshold: 1 minute, loss factor: 0.0054, annual profits: € 2500.

The niche size $\sigma_{share}= 0.1$. This value has been chosen after several runs in order to select the most appropriate set of values for the different parameters of the EA. The values of some other parameters are the following: crossover probability is 0.1; mutation probability varies between 0.0001 and 0.005; the selection operator is based on tournament in which the group for comparison has size 4.

The main characteristics of the three implementations are described in table 1.

WF denotes the implementation of the EA resorting to fitness sharing only, as a method to improve the diversity in the population. In general, it is not effective in preventing very close individuals in the population since these individuals will probably have similar values for their fitness, and being very close the fitness sharing imposes a similar degradation. In the WD implementation, indifference thresholds in each dimension are used according to the δ-non-dominated concept. WD does not use fitness sharing. In the WFD implementation both approaches are used.

Table 1. Algorithms being compared.

Simulation	Description
WF	Base case: fitness sharing only
WD	Without fitness sharing and with the indifference thresholds
WFD	With fitness sharing and with indifference thresholds

In table 2, the values *a/b* for each pair resulting from the combinations of the WF, WD and WFD algorithms are presented. According to these values it is not possible to obtain a statistically significant conclusion about the pairs WF/WD or WFD/WD, since there is no difference between the algorithms for a 95% confidence level. For example, WD is better that WF in about 6% of the lines and in 94% of times none of the algorithms performs better than the other. In the case WFD/WD, WFD is better in about 21% of the times, but there are no statistically significant differences in about 79% of the lines. On the other hand, WFD performs better that WF in about 80% of the front, while one can not say that they are different in the remaining 20% of the front.

Table 2. Comparison of the algorithms.

	1000 lines	**5000 lines**	**10000 lines**
WF/WD	0% / 5.7%	0 / 6.06%	0 / 5.65%
WFD/WF	78.3% / 0	79.52% / 0%	80.17% / 0%
WFD/WD	21.2% / 0%	21.58% / 0%	19.82% / 0%

Table 3. Comparison of all algorithms.

Algorithm	*beats all*	*unbeaten*
WFD	19.2 %	100 %
WF	0 %	4.9 %
WD	0 %	80.8 %

In table 3 the results for the statistics proposed in Knowles and Corne (2000) are presented. WFD beats WF and WD in about 20% of the front while it is unbeaten in the entire front, meaning that no other algorithm is better than WFD in any region of the front, for a 95% confidence level. WF and WD never beat the other algorithms, while they are unbeaten in about 5% of the front (WF) and 81% (WD).

According to these experiments the algorithm using both strategies (δ-non-dominance and fitness sharing) seems to produce better results, in this problem.

6 Conclusion

In the presence of search spaces whose characteristics favour the existence in the population of individuals too close, the need arises for some filtering techniques in order to avoid such situations. In this work, the so-called δ-non-dominance concept has been used. The results for a case study in load management show that the solutions obtained with the combined implementation of fitness sharing and δ-non-dominance concept are better that the ones obtained with the use of fitness sharing only. One of the most attractive characteristics of the δ-non-dominance concept is its simplicity, both regarding its implementation and the identification of the indifference thresholds in each dimension, making the process of tuning the parameters for an EA easier and quicker. This approach requires the specification of the indifference thresholds from the DM and provides him/her with solutions better spread along the Pareto optimal front.

References

Chen, J.; F. N. Lee; A. M. Breipohl; R. Adapa, "Scheduling direct load control to minimize system operation cost", IEEE Trans. on Power Systems, 10, 4, 1994-2001, 1995.

Coello, C. A. C.; D. Van Veldhuizen; Gary B. Lamont; "Evolutionary Algorithms for Solving Multi-Objective Problems"; Kluwer Academic Press; 2002.

Deb, K.; A. Pratap; S. Agarwal; T. Meyarivan, "A Fast and Elitist Multi-Objective Genetic Algorithm: NSGA-II", KanGAL Report No. 200001, 2000.

Deb, K.; "Multi-Objective Optimization using Evolutionary Algorithms"; John Wiley & Sons, Ltd, New York; 2001.

De Jong, K. A., "An analysis of the behaviour of a classe of genetic adaptive systems"; Doctoral Thesis, 1975; University of Michigan.

Fonseca, C. M.; P. J. Fleming; "On theperformance assessment and comparison of stochastic multiobjective optimizers". In Voigt, Ebeling, Rechenberg and Schwefel, editors, *Proceedings of Parallel Problem Solving from Nature IV*, 584-593, 1996.

Fonseca, C. M.; P. J. Fleming; "Multiobjective Optimization and Multiple Constraint Handling with Evolutionary Algorithms-Part I: A Unified Formulation", IEEE Trans. on Systems, Man, and Cybernetics-Part A: Systems and Humans, 28, 1, 26-37, 1998.

Goldberg, D. E.; Genetic Algorithms in Search, Optimisation and Machine Learning; Addison-Wesley; 1989.

Gomes, A., A. G. Martins, R. Figueiredo, "Simulation-based Assessment of Electric Load Management Programs", International Journal of Energy Research, 23, 169-181, 1999.

Gomes, A.; C. H. Antunes; A. G. Martins; "A multiple objective evolutionary approach for the design and selection of load control strategies"; IEEE Trans. on Power Systems, (forthcoming).

Knowles, J. D.; D. W. Corne; "Approximating the Nondominated Front Using the Pareto Archived Evolution Strategy"; Evolutionary Computation Journal, 8, 2, 149-172, 2000.

Laummans, M.; L. Thiele; K. Deb; E. Zitzler, "Combining Convergence and Diversity in Evolutionary Multi-Objective Optimization"; Evolutionary Computation; 10, 3, 2002.

Ng, K.-H.; G. B. Sheble, "Direct load control - A profit-based load management using linear programming"; IEEE Trans. on Power Systems; 13, 2, 688-695, 1998.

Osyczka, A.; "Evolutionary Algorithms for Single and Multicriteria Design Optimization"; Physica-Verlag; 2002.

Zitzler, E.; L. Thiele, "An Evolutionary Algorithm for Multiobjective Optimization: The Strength Pareto Approach", TIK Report 43; Computer Engineering and Communication Networks Lab; Swiss Federal Institute of Technology; Zurich; Switzerland, May, 1998.

Zitzler, E.; L. Thiele; K. Deb; "Comparison of Multiobjective Evolutionary Algorithms: Empirical Results", Evolutionary Computation, 8, 2, 173-195, 2000.

A Study into Ant Colony Optimisation, Evolutionary Computation and Constraint Programming on Binary Constraint Satisfaction Problems

Jano I. van Hemert[1] and Christine Solnon[2]

[1] CWI, P.O. Box 94079, NL-1090 GB Amsterdam, The Netherlands
jvhemert@cwi.nl
[2] LIRIS, CNRS FRE 2672, University of Lyon 1, Nautibus,
43 bd du 11 novembre, 69 622 Villeurbanne cedex, France
csolnon@liris.cnrs.fr

Abstract. We compare two heuristic approaches, evolutionary computation and ant colony optimisation, and a complete tree-search approach, constraint programming, for solving binary constraint satisfaction problems. We experimentally show that, if evolutionary computation is far from being able to compete with the two other approaches, ant colony optimisation nearly always succeeds in finding a solution, so that it can actually compete with constraint programming. The resampling ratio is used to provide insight into heuristic algorithms performances. Regarding efficiency, we show that if constraint programming is the fastest when instances have a low number of variables, ant colony optimisation becomes faster when increasing the number of variables.

1 Introduction

Solving constraint satisfaction problem s (CSPs) involves finding an assignment of values to variables that satisfies a set of constraints. This general problem has many real-life applications, such as time-tabling, resource allocation, pattern recognition, and machine vision.

To solve CSPs, one may explore the search space in a systematic and complete way, until either a solution is found, or the problem is proven to have no solution [1]. In order to reduce the search space, this kind of complete approach is usually combined with filtering techniques that narrow the variables domains with respect to some partial consistencies. Completeness is a very desirable feature, but it may become intractable on hard combinatorial problems. Hence, incomplete approaches, such as evolutionary computation and ant colony optimisation, have been proposed. These approaches do not perform an exhaustive search, but try to quickly find approximately optimal solutions in an opportunistic way: the search space is explored stochastically, using heuristics to guide the search towards the most-promising areas.

[2] and [3] provide experimental comparisons of different evolutionary algorithms with complete methods for solving binary constraint satisfaction problems. Both studies show that evolutionary computation is far from being able

J. Gottlieb and G.R. Raidl (Eds.): EvoCOP 2004, LNCS 3004, pp. 114–123, 2004.
© Springer-Verlag Berlin Heidelberg 2004

to compete with complete methods. A first motivation of this paper is to go on these comparative studies by integrating another heuristic approach, based on the Ant Colony Optimisation metaheuristic. Another motivation is to propose an explanation of the reasons for the differences of performances of the three approaches. Indeed, [4] introduces the notion of resampling ratio, that allows one to measure how often an algorithm re-samples the search space, i.e., generates a candidate solution already generated before. This measure is used in this paper to get insight into search efficiency of the three compared approaches.

In the next section we explain what constraint satisfaction problems are and what makes them an interesting object of study. Then in Section 3 we introduce the three algorithms involved in the empirical comparison, which is presented in Section 4. Concluding remarks are given in Section 5.

2 Constraint Satisfaction

A constraint satisfaction problem (CSP) [1] is defined by a triple (X, D, C) such that X is a finite set of variables, D is a function that maps every variable to its finite domain and C is a set of constraints. A label, denoted by $\langle X_i, v_i \rangle$, is a variable-value pair that represents the assignment of value v_i to variable X_i. An assignment, denoted by $\mathcal{A} = \{\langle X_1, v_1 \rangle, \dots, \langle X_k, v_k \rangle\}$, is a set of labels and corresponds to the simultaneous assignment of values $v_1, \dots, v_k$ to variables $X_1, \dots, X_k$ respectively. An assignment is complete if it contains a label for each variable of the CSP. A solution of a CSP is a complete assignment that satisfies all the constraints.

2.1 Random Binary CSPs

Binary CSPs only have binary constraints, that is, each constraint involves two variables exactly. Binary CSPs may be generated at random. A class of randomly generated CSPs is characterized by four components $\langle n, m, p_1, p_2 \rangle$ where n is the number of variables, m is the number of values in each variable's domain, $p_1 \in [0, 1]$ is a measure of the density of the constraints, i.e., the number of constraints, and $p_2 \in [0, 1]$ is a measure of the tightness of the constraints, i.e., the number of inconsistent pairs of values for each constraint.

Experiments reported in this paper were obtained with random binary CSPs generated according to Model E as described in [5], that is, the set of conflicting value pairs is created by uniformly, independently and with repetitions selecting $p_e \binom{n}{2} m^2$ edges out of the $\binom{n}{2} m^2$ possible ones. This process guarantees that no flawed variables are generated, which would otherwise make instances easy to solve.

After generating a problem instance with Model E, we can measure the density and the tightness of the constraints for this instance. The density of the constraints p_1 is often equal to one, except for very small (< 0.05) values of p_e. The tightness of the constraints p_2 is always smaller than or equal to p_e because of the possible repetitions. With sufficiently high p_e values (> 0.11) the p_2 value will be lower than p_e.

2.2   Phase-Transitions

When considering a class of combinatorial problems, rapid transitions in solvability may be observed as an order parameter is changed [6]. These "phase-transitions" occur when evolving from instances that are under-constrained, and therefore solved rather easily, to instances that are over-constrained, whose inconsistency is thus proven rather easily. Harder instances usually occur between these two kinds of "easy" instances, when approximately 50% of the instances are satisfiable.

In order to predict the phase-transition region, [7] introduces the notion of "constrainedness" of a class of problems, noted κ, and shows that when κ is close to 1, instances are critically constrained, and belong to the phase-transition region. For a class of random binary CSPs $\langle n, m, p_1, p_2 \rangle$, [7] defines this constrainedness by $\kappa = \frac{n-1}{2} p_1 \log_m \left(\frac{1}{1-p_2} \right)$.

One might think that phase-transitions only concern complete approaches, as they are usually associated with transitions from solvable to unsolvable instances, and incomplete approaches cannot detect whether a problem is not solvable. However, different studies (e.g., [8,9]) have shown that very similar phase-transition phenomena may also be observed with incomplete approaches.

3   Algorithms

3.1   Ant Colony Optimisation

The main idea of the Ant Colony Optimization (ACO) metaheuristic [10] is to model the problem as the search of a best path in a "construction graph" that represents the states of the problem. Artificial ants walk through this graph, looking for good paths. They communicate by laying pheromone trails on edges of the graph, and they choose their path with respect to probabilities that depend on the amount of pheromone previously left.

The ACO algorithm considered in our comparative study is called Ant-solver, and it is described in [11]; we briefly recall below the main features of this algorithm.

Construction graph and pheromone trails: The construction graph associates a vertex with each variable/value pair $\langle X_i, v \rangle$ such that $X_i \in X$ and $v \in D(X_i)$. There is a non oriented edge between any pair of vertices corresponding to two different variables. Ants lay pheromone trails on edges of the construction graph. Intuitively, the amount of pheromone laying on an edge $(\langle X_i, v \rangle, \langle X_j, w \rangle)$ represents the learned desirability of assigning simultaneously value v to variable X_i and value w to variable X_j. As proposed in [12], pheromone trails are bounded between τ_{min} and τ_{max}, and they are initialized at τ_{max}.

Construction of an assignment by an ant: At each cycle, each ant constructs an assignment, starting from an empty assignment $\mathcal{A} = \emptyset$, by iteratively adding labels to $\mathcal{A}$ until $\mathcal{A}$ is complete. At each step, to select a label, the ant first chooses a variable $X_j \in X$ that is not yet assigned in $\mathcal{A}$. This choice is performed with respect to the smallest-domain ordering heuristic, i.e., the ant selects a variable that has the smallest number of consistent values with respect to the partial assignment $\mathcal{A}$ under construction. Then, the ant chooses a value $v \in D(X_j)$ to be assigned to X_j. The choice of v is done with respect to a probability which depends on two factors, respectively weighted by two parameters α and β: a pheromone factor —which corresponds to the sum of all pheromone trails laid on all edges between $\langle X_j, v \rangle$ and the labels in $\mathcal{A}$— and a heuristic factor — which is inversely proportional to the number of new violated constraints when assigning value v to variable X_j.

Local improvement of assignments: Once a complete assignment has been constructed by an ant, it is improved by performing some local search, i.e., by iteratively changing some variable-value assignments. Different heuristics can be used to choose the variable to be repaired and the new value to be assigned to this variable. For all experiments reported below, we have used the min-conflict heuristics [13], i.e., we randomly select a variable involved in some violated constraint, and then we assign this variable with the value that minimizes the number of constraint violations.

Pheromone trails update: Once every ant has constructed an assignment, and improved it by local search, the amount of pheromone laying on each edge is updated according to the ACO metaheuristic, i.e., all pheromone trails are uniformly decreased —in order to simulate some kind of evaporation that allows ants to progressively forget worse paths— and then pheromone is added on edges participating to the construction of the best assignment of the cycle —in order to further attract ants towards the corresponding area of the search space.

3.2   Evolutionary Computation

In the most basic form, an evolutionary algorithm that solves constraint satisfaction problems uses a population of candidate solutions, i.e., complete assignments. The fitness is equal to the number of violated constraints, which needs to be minimised to zero. Such a basic approach does not yield a powerful constraint solver. To improve effectiveness and efficiency, many additional mechanisms such as heuristic operators, repair mechanisms, adaptive schemes, and different representations exist. Craenen et al. [3] have performed a large scale empirical comparison on 11 evolutionary algorithms that employ a variety of mechanisms. They conclude that the three best algorithms, Heuristics GA version 3, Stepwise Adaptation of Weights and the Glass-Box approach, have performances that are statistically not significantly different. However, all three are significantly better than the other 8 algorithms. This conclusion differs from [2] because the newer study considers even harder and larger problem instances.

We have selected the Glass-Box approach in our study and we use the same implementation as in [3], which is called JavaEa2 [14]. It can be downloaded for free and used under the GNU General Public License [15]. We provide a short overview of the algorithm to provide insight into its concept, for a more detailed description of the algorithm and all its parameters we refer to [16,3].

The Glass-Box approach is proposed by Marchiori [16]. By rewriting all constraints into one form the algorithm can then process these constraints one at a time. For each constraint that is violated the corresponding candidate solution is repaired with a local search such that this constraint no longer is violated. However, previously repaired constraints are not revisited, thus a repaired candidate solutions not necessarily yields a feasible solution. The evolutionary algorithm keeps the local search going by providing new candidate solutions. A heuristic is added to the repair mechanism by letting it start with the constraint involved in the most violations.

3.3   Constraint Programming

Constraint programming dates back to 1965 when Chronological Backtracking was proposed [17]. Since then many variants have been proposed that rely on the basic principle of using recursion to move through the search space in an exhaustive way. The main interest in creating more variants is to speed up the process by making smarter moves and decisions. These involve a better order in which the variables and their domains are evaluated, as well as providing mechanisms that make it possible to take larger leaps through the search tree.

Here we use a constraint programming algorithm that uses forward checking (FC) of Haralick and Elliot [18] for its constraint propagation, which means that the algorithm first assigns a value to a variable and then checks all unassigned variables to see if a valid solution is still possible. The goal of forward checking is to prune the search tree by detecting inconsistencies as soon as possible. To improve upon the speed, conflict-directed backjumping (CBJ) [19] was added to it by Prosser [20] to form FC-CBJ. The speed improvement comes from the ability to make larger jumps backward when the algorithm gets into a state where it needs to backtrack to a previous set of assignments by jumping back over variables of which the algorithm knows that these will provide no solutions.

4   Experiments

This section reports experimental results obtained by the three algorithms Antsolver, Glass-Box, and FC-CBJ, in order to compare their effectiveness and efficiency.

To determine the effectiveness of the algorithms, we compare their success ratio, i.e., the percentage of runs that have succeeded in finding a solution. Note that, without any time limit, complete approaches always succeed in finding a solution as we only consider solvable instances.

To determine the efficiency of the algorithms, we compare the number of conflict checks they have performed. A conflict check is defined as querying if the assignment of two variables is allowed. Thereby we can characterize the speed of the algorithms independently from implementation and hardware issues as all algorithms have to perform conflict checks, which take the most time in execution.

Finally, the resampling ratio is used to get insight into how efficient algorithms are in sampling the search space. If we define S as the set of unique candidate solutions generated by an algorithm over a whole run and evals as the total number of generated candidate solutions then the resampling ratio is defined as

$\frac{evals-|S|}{evals}$. Low values correspond with an efficient search, i.e., not many duplicate candidate solutions are generated. Note that complete approaches never generate twice a same candidate solution, so that their resampling ratio is always zero.

4.1 Experimental Setup

The Ant-solver algorithm has been run with the following parameters setting: $\alpha = 2, \beta = 10, \rho = 0.99, \tau_{min} = 0.01, \tau_{max} = 4$, as suggested in [11]. The number of ants has been set to 15 and the maximum number of cycles to 2 000, so that the number of generated candidate solutions is limited to 30 000.

The Glass-Box algorithm has been run with a population size of 10, a generational model using linear ranking selection with a bias of 1.5 to determine the set of parents. The mutation rate was set to 0.1. The evolutionary algorithm terminates when either a solution is found or the maximum of 100 000 evaluations is reached.

For both Ant-solver and Glass-Box, we have performed 10 independent runs for each instance, and we report average results over these 10 runs.

4.2 Test Suites

We compare the three algorithms on two test suites of instances, all generated according to Model E.

Test suite 1 is used to study what happens when moving towards the phase transition region —where the most difficult instances are supposed to be. It contains 250 instances, all of them having 20 variables and 20 values in each variable's domain, i.e., $n = m = 20$, and a density of constraints p_1 equals to 1, i.e., there is a constraint between any pair of different variables. These 250 instances are grouped into 10 sets of 25 problem instances per value of the tightness parameter p_2, ranging between 0.21 and 0.28, as the phase transition region occurs when $p_2 = 0.266$ (which corresponds with $p_e = 0.31$ when using Model E).

Test suite 2 is used to study the scale-up properties of the algorithms within the phase transition region. It is composed of 250 instances, grouped into 10 sets of 25 instances per number of variables n, ranging from 15 to 60. For all these instances, the domain size m of each variable is kept constant at 5. To make sure the problem instances have similar positions with respect to the phase-transition region, so that the difficulty of instances only depends on the number of variables, we keep the constrainedness $\kappa = 0.92$ on average (with a standard deviation of 0.019) by setting appropriately the order parameter p_e.

4.3 Moving towards the Phase-Transition Region

Table 1 reports results obtained by each algorithm for each group of 25 instances of test suite 1. This table clearly shows that for the Glass-Box approach, one of the best known evolutionary algorithms for solving binary CSPs, the success ratio quickly drops from one to almost zero when moving towards the phase transition. Set against this result, Ant-solver seems a much better competition for the complete method FC-CBJ, being able to find a solution for nearly all the runs.

Table 1. Results of Glass-Box (GB), Ant-solver (AS), and FC-CBJ on the 250 $\langle 20, 20, 1, p_2 \rangle$ instances of test suite 1, grouped into 10 sets of 25 instances per value of p_2 ranging between 0.213 and 0.280.

p_2	0.213	0.221	0.228	0.236	0.244	0.251	0.259	0.266	0.273	0.280
	Success ratio									
GB	1.00	0.97	0.86	0.65	0.36	0.17	0.10	0.02	0.03	0.02
AS	1.00	1.00	1.00	1.00	1.00	1.00	1.00	0.96	0.98	1.00
FC-CBJ	1.00	1.00	1.00	1.00	1.00	1.00	1.00	1.00	1.00	1.00
	Resampling ratio									
GB	0.10	0.23	0.40	0.60	0.78	0.87	0.90	0.94	0.93	0.93
AS	0.00	0.00	0.00	0.00	0.00	0.00	0.01	0.14	0.13	0.08
FC-CBJ	0.00	0.00	0.00	0.00	0.00	0.00	0.00	0.00	0.00	0.00
	Average number of conflict checks									
GB	2.4e6	1.5e7	5.3e7	1.2e8	2.0e8	2.6e8	3.1e8	3.6e8	3.9e8	4.3e8
AS	2.5e5	4.4e5	1.0e6	2.5e6	6.3e6	1.8e7	5.4e7	1.2e8	1.1e8	9.0e7
FC-CBJ	5.7e4	1.2e5	2.1e5	7.0e5	1.7e6	5.2e6	2.0e7	4.1e7	4.1e7	2.6e7
	Standard deviation of the number of conflict checks									
GB	1.4e7	5.3e7	1.0e8	1.3e8	1.3e8	1.1e8	1.1e8	8.0e7	9.3e7	9.8e7
AS	2.4e5	4.7e5	9.6e5	2.4e6	6.6e6	1.8e7	5.2e7	2.8e8	2.0e8	1.1e8
FC-CBJ	5.3e4	1.5e5	2.9e5	7.5e5	1.4e6	4.2e6	2.9e7	3.1e7	2.4e7	1.5e7
	Constraint checks ratio of GB and AS w.r.t. FC-CBJ									
GB/FC-CBJ	42.11	125.00	252.38	171.43	117.65	50.00	15.50	8.78	9.51	16.54
AS/FC-CBJ	4.39	3.67	4.76	3.57	3.71	3.46	2.70	2.93	2.68	3.46

The resampling ratio reveals what happens. As the success ratio of Glass-Box drops, its resampling ratio keeps increasing, even to levels of over 90%, indicating that Glass-Box does not diversify enough its search. For Ant-solver we only observe a slight increase of the resampling ratio, up to 14% around the most difficult point at $p_2 = 0.266$, which corresponds with a slight drop in its success ratio.

However, when we compare the efficiency of the three algorithms by means of the number of conflict checks, we notice that although Ant-solver is faster than Glass-Box, it is slower than FC-CBJ. The last two lines of table 1 display the ratio of the number of conflict checks performed by Glass-Box and Ant-solver with respect to FC-CBJ, and show that Glass-Box is from 8 to 250 times as slow as FC-CBJ whereas Ant-solver is from two to five times as slow than FC-CBJ. Note that for Ant-solver this ratio is rather constant: for Ant-solver and FC-CBJ, the number of conflict checks increases in a very similar way when getting closer to the phase transition region.

4.4   Varying the Number of Variables

Table 2 reports results obtained by the three algorithms for each group of 25 instances of test suite 2. In this table, one can first note that the effectiveness of the Glass-Box approach decreases when increasing the number of variables and although the rise in efficiency seems to slow down, this corresponds with success ratios less than 50%. The resampling ratio shows inverse behaviour with the success ratio. On the contrary, Ant-solver always finds a solution, while its resampling ratio is always less than 3% for each setting. The search of Ant-solver is thus extremely efficient for this scale-up experiment.

Table 2. Results of Glass-Box (GB), Ant-solver (AS), and FC-CBJ on the 250 $\langle n, 5, p_1, p_2 \rangle$ instances of test suite 2, grouped into 10 sets of 25 instances per value of n ranging between 15 and 60.

n	15	20	25	30	35	40	45	50	55	60
p_1	0.99	0.98	0.95	0.92	0.89	0.85	0.81	0.78	0.74	0.72
p_2	0.19	0.15	0.12	0.10	0.093	0.086	0.080	0.075	0.071	0.069
	Success ratio									
GB	0.86	0.70	0.58	0.57	0.38	0.28	0.23	0.20	0.11	0.14
AS	1.00	1.00	1.00	1.00	1.00	1.00	1.00	1.00	1.00	1.00
FC-CBJ	1.00	1.00	1.00	1.00	1.00	1.00	1.00	1.00	1.00	1.00
	Resampling ratio									
GB	0.32	0.47	0.51	0.51	0.58	0.64	0.65	0.63	0.62	0.61
AS	0.021	0.014	0.008	0.003	0.002	0.001	0.001	0.000	0.000	0.000
FC-CBJ	0.00	0.00	0.00	0.00	0.00	0.00	0.00	0.00	0.00	0.00
	Average number of conflict checks									
GB	1.1e7	3.6e7	7.1e7	1.0e8	1.9e8	2.7e8	3.7e8	4.4e8	6.0e8	7.0e8
AS	6.0e4	2.4e5	7.0e5	1.5e6	4.7e6	8.4e6	1.6e7	4.0e7	6.5e7	1.0e8
FC-CBJ	5.1e3	3.9e4	2.0e5	8.9e5	4.7e6	3.2e7	7.8e7	5.3e8	3.0e9	9.0e9
	Standard deviation of the number of conflict checks									
GB	2.3e7	4.8e7	7.7e7	1.1e8	1.4e8	1.6e8	1.8e8	2.1e8	2.0e8	2.7e8
AS	6.8e4	2.8e5	8.5e5	1.8e6	6.7e6	9.3e6	2.0e7	4.7e7	6.2e7	7.6e7
FC-CBJ	4.4e3	3.3e4	1.8e5	7.2e5	3.4e6	4.0e7	6.9e7	4.6e8	3.2e9	7.6e9
	Conflict checks ratio of GB and AS w.r.t. FC-CBJ									
GB/FC-CBJ	2156.86	923.08	355.00	112.36	40.43	8.44	4.74	0.83	0.20	0.08
AS/FC-CBJ	11.76	6.15	3.50	1.69	1.00	0.26	0.21	0.08	0.02	0.01

Regarding efficiency, we note that FC-CBJ starts out as the fastest algorithm at first, but at 35 variables it loses this position to Ant-solver. This difference in efficiency is clearly visible in the last line of Table 2: for instances with 15 variables, FC-CBJ is 12 times as fast than Ant-solver, whereas for instances with 60 variables, it is 100 times as slow.

Actually, the number of conflict checks performed by Ant-solver for computing one candidate solution is in $\mathcal{O}(n^2 m)$ (where n is the number of variables and

m the size of the domain of each variable), and the number of computed candidate solutions is always bounded to 30 000. Hence, the complexity of Ant-solver grows in a quadratic way with respect to the number of variables, and experimental results of Table 2 actually confirm this. On the contrary, the theoretical complexity of tree-search approaches such as FC-CBJ grows exponentially with respect to the number of variables. Even though FC-CBJ uses elaborate backtracking techniques, i.e., trying to jump over unsuccessful branches of the search tree, experimental results show us that the number of conflict checks performed by FC-CBJ still increases exponentially.

5   Conclusions

Experiments reported in this paper showed us that, considering effectiveness, ant colony optimisation is a worthy competitor to constraint programming, as Ant-solver almost always finds a solution, whereas evolutionary computation more often fails in finding solutions when getting closer to the phase transition region, or when increasing the size of the instances.

Scale-up experiments showed us that, on rather small instances, the constraint programming approach FC-CBJ is much faster than the two considered incomplete approaches. However, run times of FC-CBJ grow exponentially when increasing the number of variables, so that on larger instances with more than 35 variables, its efficiency becomes significantly lower than Ant-solver and Glass-Box.

The resampling ratio is a good indication of the problems that occur during the search in both the ant colony optimisation and the evolutionary algorithms. Whenever an algorithm's resampling increases this corresponds to a decrease in effectiveness. At the same time the efficiency will also drop.

Future research will incorporate experiments on over constrained instances, the goal of which is to find an assignment that maximises the number of satisfied constraints.

Acknowledgements. We thank Bart Craenen for developing the JavaEa2 platform and making it freely available to the community.

References

1. Tsang, E.: Foundations of Constraint Satisfaction. Academic Press (1993)
2. van Hemert, J.: Comparing classical methods for solving binary constraint satisfaction problems with state of the art evolutionary computation. In Cagnoni, S., Gottlieb, J., Hart, E., Middendorf, M., Raidl, G., eds.: Applications of Evolutionary Computing. Number 2279 in Springer Lecture Notes on Computer Science, Springer-Verlag, Berlin (2002) 81–90
3. Craenen, B., Eiben, A., van Hemert, J.: Comparing evolutionary algorithms on binary constraint satisfaction problems. IEEE Transactions on Evolutionary Computation **7** (2003) 424–444

4. van Hemert, J., Bäck, T.: Measuring the searched space to guide efficiency: The principle and evidence on constraint satisfaction. In Merelo, J., Panagiotis, A., Beyer, H.G., Fernández-Villacañas, J.L., Schwefel, H.P., eds.: Parallel Problem Solving from Nature — PPSN VII. Number 2439 in Springer Lecture Notes on Computer Science, Springer-Verlag, Berlin (2002) 23–32

5. Achlioptas, D., Kirousis, L., Kranakis, E., Krizanc, D., Molloy, M., Stamatiou, Y.: Random constraint satisfaction: A more accurate picture. Constraints **4** (2001) 329–344

6. Cheeseman, P., Kenefsky, B., Taylor, W.M.: Where the really hard problems are. In: Proceedings of the IJCAI'91. (1991) 331–337

7. Gent, I., MacIntyre, E., Prosser, P., Walsh, T.: The constrainedness of search. In: Proceedings of AAAI-96, AAAI Press, Menlo Park, California. (1996)

8. Davenport, A.: A comparison of complete and incomplete algorithms in the easy and hard regions. In: Proceedings of CP'95 workshop on Studying and Solving Really Hard Problems. (1995) 43–51

9. Clark, D., Frank, J., Gent, I., MacIntyre, E., Tomv, N., Walsh, T.: Local search and the number of solutions. In: Proceedings of CP'96, LNCS 1118, Springer Verlag, Berlin, Germany. (1996) 119–133

10. Dorigo, M., Caro, G.D., Gambardella, L.M.: Ant algorithms for discrete optimization. Artificial Life **5** (1999) 137–172

11. Solnon, C.: Ants can solve constraint satisfaction problems. IEEE Transactions on Evolutionary Computation **6** (2002) 347–357

12. Stützle, T., Hoos, H.: $\mathcal{MAX} - \mathcal{MIN}$ Ant System. Journal of Future Generation Computer Systems **16** (2000) 889–914

13. Minton, S., Johnston, M., Philips, A., Laird, P.: Minimizing conflicts: a heuristic repair method for constraint satistaction and scheduling problems. Artificial Intelligence **58** (1992) 161–205

14. Craenen, B.: JaveEa2: an evolutionary algorithm experimentation platform for constraint satisfaction in Java (Version 1.0.1) `http://www.xs4all.nl/~bcraenen/JavaEa2/download.html`.

15. Foundation, F.S.: The GNU general public license (Version 2, June 1991) `http://www.gnu.org/licenses/gpl.txt`.

16. Marchiori, E.: Combining constraint processing and genetic algorithms for constraint satisfaction problems. In Bäck, T., ed.: Proceedings of the 7th International Conference on Genetic Algorithms, Morgan Kaufmann (1997) 330–337

17. Golomb, S., Baumert, L.: Backtrack programming. ACM **12** (1965) 516–524

18. Haralick, R., Elliot, G.: Increasing tree search efficiency for constraint-satisfaction problems. Artificial Intelligence **14** (1980) 263–313

19. Dechter, R.: Enhancement schemes for constraint processing: Backjumping, learning, and cutset decomposition. Artificial Intelligence **41** (1990) 273–312

20. Prosser, P.: Hybrid algorithms for the constraint satisfaction problem. Computational Intelligence **9** (1993) 268–299

Binary Merge Model Representation of the Graph Colouring Problem

István Juhos[1], Attila Tóth[2], and Jano I. van Hemert[3]

[1] Dept. of Computer Algorithms and Artificial Intelligence, University of Szeged
[2] Department of Computer Science, University of Szeged (JGYTFK)
[3] National Institute for Mathematics and Computer Science, Amsterdam (CWI)

Abstract. This paper describes a novel representation and ordering model that, aided by an evolutionary algorithm, is used in solving the graph k-colouring problem. Its strength lies in reducing the number of neighbors that need to be checked for validity. An empirical comparison is made with two other algorithms on a popular selection of problem instances and on a suite of instances in the phase transition. The new representation in combination with a heuristic mutation operator shows promising results.

1 Introduction

Constraint satisfaction problems form a well studied class of problems [1]. Within this class the graph k-colouring problem is one of the most popular problems. The second DIMACS challenge [2] has been devoted to it, which has led to a standard file format for representing problem instances. Here we will use this file format, together with problem instances from the challenge.

In this paper we introduce a novel approach to representing candidate solutions for the graph k-colouring problem in an evolutionary algorithm. The approach is based on effective reduction rules. The order in which these reduction rules are executed determines the quality of the solution. This order will be determined by an evolutionary algorithm.

To show that the approach based on reduction rules has a performance that is competitive with other approaches, we perform a comparison with two algorithms that are known to perform well. The comparison is based on the effectiveness and efficiency of the algorithms.

The structure of the remainder of the paper is as follows. In the next section we give a definition and short overview of graph k-colouring. In Section 3, we introduce the new representation. This will be incorporated into an evolutionary algorithm described in Section 4, which we will compare empirically to other methods in Section 5. In Section 6 we provide conclusions.

2 Graph k-Colouring

The problem class known as graph k-colouring is defined as follows, given a graph $\langle V, E \rangle$ where $V = \{v_1, ..., v_n\}$ is a set of nodes and $E = \{(v_i, v_j) | v_i \in V \wedge v_j \in$

J. Gottlieb and G.R. Raidl (Eds.): EvoCOP 2004, LNCS 3004, pp. 124–134, 2004.

$V \wedge i \neq j\}$ is a set of edges, where every edge lies between two nodes. The graph k-colouring problem is to colour every node in V with one of k colours such that no two nodes connected with an edge in E have the same colour.

Many algorithms to solve this problem were created and studied. Early algorithms are complete, i.e., they always find a solution if one exists. However, it was proven that such algorithms exhibit an exponential effort to either find a solution or proof that no solution exists, when the problem is scaled up. As a result many algorithms studied today use a stochastic process to guide them towards suboptimal solutions or, hopefully, towards an optimal solution. Examples of such methods include Tabu-search [3], simulated annealing [4] and ant colony optimisation [5].

One popular approach for dealing with graph k-colouring is evolutionary computation [6] Unfortunately, evolutionary algorithms are not necessarily very good in solving constraint satisfaction problems [7] as they may suffer from a number of flaws that keeps them from reaching optimal solutions [8]. One problem that is appropriate for graph k-colouring is when the problem at hand contains symmetry [9]. For the latter, symmetry is defined as the invariance to the solution when permuting he colours. Besides symmetry, other structural properties make it difficult to find solutions efficiently. As we know that graph k-colouring exhibits a phase transition depending on the ratio of constraints to the maximum number of possible constraints $\binom{n}{2}$ [10], where the number of solvable problem instances quickly drops to zero. At that transition constraint solvers require the most search effort to find solutions for solvable problem instances. Recent investigations [11] have shown that the size of the backbone in graph k-colouring may be a good explanation towards explaining the rise in difficulty at the phase transition.

3   Representing the Graph k-Colouring Problem

3.1   Binary Merge Model

Graph colouring algorithms make use of adjacency checking during colouring, which plays a significant role in performance. The number of checks depends on the representation of the problem. The model introduced here directly addresses this issue. Generally, when assigning a colour to a node all adjacent nodes must be scanned to check for an equal colouring, i.e., constraint check needs to be performed. Thus, we have to perform at least as many checks as the number of coloured neighbours and at most as many as $|V| - 1$. In the new model, this number of checks is between one and the number of colours used up to this point.

The Binary Merge Model (BMM) implicitly uses hyper-nodes and hyper-edges (see Figure 1(b)). A hyper-edge connects a hyper-node with other nodes, regardless whether it is normal or hyper. A hyper-node is a set nodes that have the same colour. If h is a hyper-node and v is a normal node that are connected by a hyper-edge then, and only then, h contains at least one normal node connected to v. Our algorithm concentrates on the relation between hyper-nodes and normal nodes. Thus, the adjacency checks can be done along hyper-edges instead of edges, whereby we spare many checks because during colouring, the initial set of edges is folded into hyper-edges.

The colouring of a graph is stored in a Binary Merge Table (BMT) (see Figure 2). Every cell (i, j) in this table takes as value either 0 or 1. The columns refer to the nodes and the rows refer to the colours. The value in cell (i, j) is 1 if and only if node j cannot be assigned colour i because of the edges in the original graph $\langle V, E \rangle$. In the initial colouring every node is assigned a unique colour by setting the (i, i) cells to 0.

If the graph is not a full graph, then it might be possible to decrease the number of necessary colours. This amounts to reducing rows in the BMT. To make reduction of the rows possible we introduce a Binary Merge Operation, which attempts to merge two rows whereby reducing the number of colours with one. It is successful only when the two nodes are not connected by an edge or by a hyper-edge. An example is found in Figure 1(b) and the related Figure 2.

Definition 1. Binary Merge Operation $\cup$ merges an initial row i to an arbitrary (initial or merged) row r if and only if $(r, i) = 0$ (i.e., the hyper-node r is not connected to the node i) in BMT. If rows i and r can be merged then the result is the union of i and r.

Formally, let I be the set of initial rows of BMT and R be the set of all possible $|V|$ size binary rows (vectors), then a binary merge operation is defined as,

$$\cup : R \times I \rightarrow R$$
$$r' := r \cup i, \text{ where } r', r \in R, i \in I, \text{ or by components}$$
$$r'_k := r_k \vee i_k, k = 1, \dots, |V|, \text{ where } \vee \text{ is the logical OR operator}$$

Regarding the time complexity of this operation, we can say that it is proportional to a binary OR operation on a register of l bits. If l is the number of bits in one such operation and under assumption that the time complexity of that operation is one, the merge of two rows of length n by l length parts takes $\lceil n/l \rceil$ to complete. If m is the number of rows left in MT, then $n - m$ binary merge operations are performed, where $m \in \{\chi, \dots, n\}$.

3.2   Permutation Binary Merge Model

The result, i.e., the colouring of a graph, after two or more binary merge operations, depends on the order in which these operations were performed. Consider the hexagon in Figure 1(a). In Figure 2 the BMTs are shown that correspond to the graphs in Figure 1. Now let the sequence of the rows 1, 4, 2, 5, 3, 6 be the order in which rows are considered for the binary merge operation and consider the following merging procedure. Take the first two rows in the sequence, then attempt to merge row 4 with row 1. As these can be merged the result is $1 \cup 4$ (see Figure 1(b)). Now take row 2 and try to merge this with the first row, i.e., $(1 \cup 4)$. This is unsuccessful, so row 2 remains unaltered. The merge operations continues with the next rows 5, 3, and finally, 6. The allowed merges are $1 \cup 4$,

$2 \cup 5$, and $3 \cup 6$. This sequence of merge operations results in the 3-colouring of the graph depicted in Figure 1(c). However, if we use the sequence $1, 3, 5, 2, 4, 6$ then the result is the 2-colouring in Figure 1(d) with the merges $1 \cup 3$, $(1 \cup 3) \cup 5$, $2 \cup 4$, and $(2 \cup 4) \cup 6$. The defined merge is greedy, it takes a row and tries to find the first row from the top of the table that it can merge. The row remains unaltered if there is no suitable row. After performing the sequence P of merge operations, we call the resulting BMT the m erged BMT.

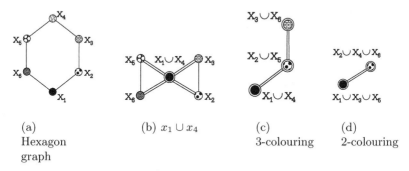

(a)
Hexagon
graph

(b) $x_1 \cup x_4$

(c)
3-colouring

(d)
2-colouring

Fig. 1. Example of the result of two different merge orders. Double-lined edges are hyper-edges and double-lined nodes are hyper-nodes.

(a)	x_1	x_2	x_3	x_4	x_5	x_6
r_1	0	1	0	0	0	1
r_2	1	0	1	0	0	0
r_3	0	1	0	1	0	0
r_4	0	0	1	0	1	0
r_5	0	0	0	1	0	1
r_6	1	0	0	0	1	0

(b)	x_1	x_2	x_3	x_4	x_5	x_6
$r_1 \cup r_4$	0	1	1	0	1	1
r_2	1	0	1	0	0	0
r_3	0	1	0	1	0	0
r_5	0	0	0	1	0	1
r_6	1	0	0	0	1	0

(c)	x_1	x_2	x_3	x_4	x_5	x_6
$r_1 \cup r_4$	0	1	1	0	1	1
$r_2 \cup r_5$	1	0	1	1	0	1
$r_3 \cup r_6$	1	1	0	1	1	0

(d)	x_1	x_2	x_3	x_4	x_5	x_6
$r_1 \cup r_3 \cup r_5$	0	1	0	1	0	1
$r_2 \cup r_4 \cup r_6$	1	0	1	0	1	0

Fig. 2. The Binary Merge Tables that correspond to the graphs in Figure 1

Finding a minimal colouring for a graph k-colouring problem using the BMT representation and binary merge operations comes down to finding the sequence of merge operations that leads to that colouring. This can be represented as a sequence of candidate reduction steps using the greedy approach described above. The permutations of this representation form the Permutation Binary Merge Model. We proof that this model is a valid representation of the graph k-colouring problem, i.e., that any valid k-colouring can be represented with it.

Lemma 1 (Equivalence). An arbitrary valid k-colburing of a graph G can be associated w ith a m erged B inary M erge Table.

Proof (constructive). Consider a $k \times n$ BMT, where n is the number of the nodes and k is the number of colours used in the colouring. Let a cell $(c, i) = 1$ if and only if node i has an adjacent node that is coloured with c, otherwise let it contain 0.

Lemma 2 (Solvability). If O is an optimal colouring of the graph G, i.e., it uses the k colours, then there exists a sequence of merge operations P that reduces the initial Binary Merge Table such that O is obtained by P.

Proof (constructive). Suppose O is an optimal solution. Construct a merged BMT B according to O, this exists because of Lemma 1. The binary merge operation is reversible if we keep track of the merges. Let this reverse operation be called "unmerge". Unmerge all merged rows in B until the number of rows is equal to the number of nodes of G. At first, apply unmerge to the last row of B as many times as it contains nodes. Then, continue to unmerge the previous row and so on. By applying the unmerge operation sequences we get a sequence of merge operators P that leads back to O. Note that there may exist several sequence that lead to O because unmerge can be applied in any optional order.

4   Evolutionary Algorithm to Guide PBMM

The search space of the model in Section 3.2 is of size $n!$ and contains many local optima. We use an evolutionary algorithm (PBMM-EA)[1] to search through the space of permutations. The genotype is thus the permutations of the rows. The phenotype is a valid colouring of the graph after using the greedy method on the permutation to select the order of binary merge operations.

An intuitive way of measuring the quality of an individual p in the population is by counting the number of rows remaining in the final BMT. This equals to the number of colours $k(p)$ used in the colouring of the graph, which needs to be minimised. If we know that the optimal colouring is χ then we may normalise this fitness function to $g(p) = k(p) - \chi$.

This function gives a rather low diversity of the individuals in a population because it cannot distinguish between two individuals that use an equal number of colours. We address this problem by introducing a new multiplier. This multiplier is based on the heuristic that we want to get rid of highly constraint rows in order to have more chance of successful merges later. This involves merges of rows where many 1s are merged. Let $z(p)$ denote the number of 1s in the final BMT B then the fitness function becomes $f(p) = (k(p) - \chi)z(p)$.

We use a steady-state evolutionary algorithm with 2-tournament selection, which incorporates elitism. The stop condition is that either an individual p exists with $f(p) = 0$ or that the maximum number of 3 000 generations is reached. The latter means that the run was unsuccessful, i.e., the optimal colouring was not found. The population size depends on the problem and will be set in the experiments.

---

[1] PBMM-EA is developed using the EO [12] framework

Two variation operators are used to provide offspring. First, the 1-point order-based crossover (OX1) [13, in Section C3.3.3.1] is applied. Second, the other variation operator is a mutation operator. We test two different mutation operators yielding two variants of the evolutionary algorithm. The probability of crossover depends on the choice of mutation operator. During tuning, we tried probabilities from 0 to 1 in several combinations. In case of DGS mutation, the results showed that a higher probability of mutation and a lower probability of crossover lead to better solutions.

PBMM-EA/SWAP. This variant first applies with a probability of 0.6 OX1 and then with a probability of 0.3 the simple swap mutation, which selects at random two different items in the permutation and then swaps them.

PBMM-EA/DGS D ifficulty G uided Swap. This variant applies with a probability of 0.3 OX1 and then always applies a heuristic mutation operator that is similar to the simple swap mutation, but it always chooses a node related to the last merged row and forces it to be ahead of its position in the permutation. To accomplish this it chooses at random a previous row identifier for swap. The idea being that these last merged rows are the most difficult to merge.

5   Empirical Comparison

5.1   Benchmark Algorithms

Stepwise Adaptation of Weights SAW is put forth in [6] as a very promising technique for colouring graph 3-colouring problems by showing its competitiveness against an early version of DSATUR and the Grouping Genetic Algorithm by Falkenauer [14]. We provide a brief overview of the concept of SAW and refer to [6] for a detailed description of the algorithm and specific versions of it.

The basic idea behind SAW is to learn on-line about the difficulty of constraints in a problem instance. This is achieved by keeping a vector of weights that associates the weight w_i with constraint i. In the context of graph k-colouring every edge $i \in E$ is assigned a weight w_i. These weights are initialised as one. A basic evolutionary algorithm is used to solve a given problem instance. Every T generations it is interrupted in order to change the vector of weights using the best individual of the current population. Every constraint i violated by this individual is incremented by one. Then, the evolutionary algorithm continues using this new vector. The fitness of an individual equals the sum of the weights of all the constraints it violates. By adapting this fitness function using the vector of weights, a dynamic process occurs that may prevent the evolutionary algorithm to get stuck in local optima. The SAW-EA is the same version as described in [6].

CLQ-DSATUR is an improvement [15] over the original DSATUR, which originates from Brélaz [16]. It uses a heuristic to dynamically change the ordering of the nodes and then applies the greedy method to colour the nodes. It works as follows, one node with the highest saturation degree, i.e., number of adjacent

colours, is selected from the uncoloured subgraph and it is assigned the lowest indexed colour that still yields a valid colouring. If there exists more than one most
saturated nodes, the algorithm chooses a node with the highest degree among
them. At first CLQ-DSATUR tries to find the largest clique, i.e., the subgraph that
is a full-graph, in the graph. It allocates different colours for the nodes in the
clique. Then, uncoloured nodes are dynamically ordered by saturation of colour
and subproblems are created as in the original DSATUR algorithm. This uses
backtracking to discover a valid colouring. More details about this algorithm is
found in [17].

5.2   Means of Comparisons

The performance of an algorithm is expressed in its effectiveness and its efficiency
in solving a problem instance. The first is measured using the success ratio, which
is the amount of runs where an algorithm has found the optimum divided by
the total number of runs. The second is measured by keeping track of how many
constraint checks are being performed on average, for a successful run. This
measure is independent of hardware and programming language as it counts the
number of times an algorithm requests information about the problem instance,
e.g., it checks if an edge exists between two nodes in the graph. This check, or
rather the number of times it is performed, forms the largest amount of time
spend by any constraint solver. A constraint check is defined for each algorithm
as checking whether the colouring of two nodes is allowed (satisfied) or not
allowed (violated).

The two variants of PBMM-EA and the SAW-EA are both stochastic algorithms.
Therefore we perform 10 independent runs with different random seeds for each
problem instance. The number of constraint checks are averaged over these 10
runs. The complete method CLQ-DSATUR suffices with 1 run.

5.3   DIMACS Challenge

We would like to present the performance of the PBMM-EA variants on different kinds of real life problem instances. The test suite consists from problem
instances taken from "The Second DIMACS Challenge" [2] and Trick's graph
colouring repository [18]. In Table 1 we show the results for 34 problem instances.

The second goal is to test the supposed benefit of the difficulty guided mutation operator over the simple swap mutation. For all runs we set the population
size for the PBMM-EA variants to two, but for the queens instances this gives
very poor results due to the equal degree in the graph, which is also visible for
the benchmark algorithm. Consequently we increase the population size to 12
for queen6_6 and to 70 for queen7_7 and queen8_8 to show the difference on such
graph structures. This sets a guideline for the next test suite.

For large graphs the PBMM-EA variants are much faster than the CLQ-DSATUR
and SAW-EA. The two PBMM-EA variants do not differ significantly except for
the miles and queens graphs where the difficulty guided swap variant shows

Table 1. Average number of constraint checks required for solving various problem instances using CLQ-DSATUR, SAW-EA, PBMM-EA/SWAP and PBMM-EA/DGS. Entries with "–" refer to where the algorithm never found the chromatic number, in all other cases the success ratio is one. The last three entries are used to show the difference in the two mutation operators.

Graph	$|V|$	$|E|$	χ	CLQ-DSATUR	SAW	PBMM-EA SWAP	DGS
fpsol2.i.1	496	11 654	65	1 311 037	49 028	13 831	17 793
fpsol2.i.2	451	8 691	30	1 078 861	1 745 560	12 196	11 513
fpsol2.i.3	425	8 688	30	973 200	1 414 220	69 964	10 276
inithx.i.1	864	18 707	54	3 640 097	155 996	15 552	21 490
inithx.i.2	645	13 979	31	2 328 921	739 201	8 447	12 475
inithx.i.3	621	13 969	31	2 150 844	304 910	9 725	12 135
mulsol.i.1	197	3 925	49	292 942	6 265	5 964	8 525
mulsol.i.2	188	3 885	31	229 907	21 707	4 110	5 667
mulsol.i.3	184	3 916	31	224 934	51 042	4 874	5 619
mulsol.i.4	185	3 946	31	228 474	128 130	4 084	5 606
mulsol.i.5	186	3 973	31	232 835	11 120	4 141	5 536
zeroin.i.1	211	4 100	49	289 622	13 165	6 670	8 040
zeroin.i.2	211	3 541	30	289 191	65 053	11 870	4 942
zeroin.i.3	206	3 540	30	277 504	52 493	22 556	11 197
anna	138	493	11	89 024	15 579	2 903	1 242
david	87	406	11	34 557	56 872	9 957	2 493
homer	561	1 629	13	1 484 278	11 906 400	71 572	5 038
huck	74	301	11	25 783	1 210	788	1 015
jean	80	254	10	29 939	11 390	746	949
miles250	128	387	8	76 961	983 894	21 556	3 051
miles500	128	1 170	20	89 025	9 724 950	191 011	20 398
miles750	128	2 113	31	136 866	7 922 930	946 683	103 376
miles1000	128	3 216	42	462 986	15 476 000	1 551 235	164 312
miles1500	128	5 198	73	404 051	886 155	167 487	67 721
myciel3	11	20	4	798	32	49	73
myciel4	23	71	5	4 273	135	120	186
myciel5	47	236	6	22 104	447	265	447
myciel6	95	755	7	117 163	5 920	708	955
myciel7	191	2 360	8	638 536	52 997	4 074	3 245
games120	120	638	9	101 070	3 227	1 492	1 926
queen5_5	25	160	5	8 797	8 835	4 630	2 000
queen6_6	36	290	7	202 170	–	740 550	139, 711
queen7_7	49	476	7	1 015 209	3 195 320	6 326 912[1]	744 273
queen8_8	64	728	9	57 989 844	–	30 412 779[2]	9 558 255

[1] success ratio of 0.9
[2] success ratio of 0.4

its improvement over the simple swap mutation. Also, the latter is not able to always find a solution for two of the queen graphs.

5.4 Phase Transition

To test the algorithms on really difficult graphs, we generate, using Culberson's generator [10], a test suite of 3-colourable graphs with 200 nodes that show a phase transition. The graphs are equipartite, i.e., for each colour the same amount of nodes, to some approximation, will be coloured. It consists of 9 groups of graphs where each group has 25 instances. The connectivity is changed from 0.020 to 0.060 by steps of 0.005 over the groups. The population size for PBMM-EA/DGS is set to 100 for these difficult graphs.

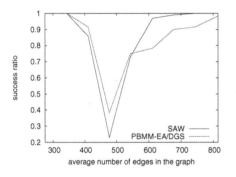 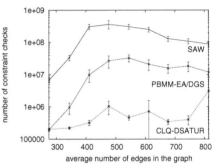

(a) Success ratio, note that CLQ-DSATUR always finds a solution

(b) Average number of constraint checks with 95% confidence intervals

Fig. 3. Results for 225 random equipartite graph 3-colouring problems of size 200 where for each problem instance 10 independent runs are performed

In Figure 3 we show the results for SAW-EA, PBMM-EA/DGS and CLQ-DSATUR. The latter is clearly the most useful algorithm as it always finds a solution and it takes the least number of constraint checks to do so. The results for the three algorithms in Figure 3(b) are significantly different and allow us to make a clear ranking on efficiency for the three algorithms.

Both evolutionary algorithms show a sharp dip in the success ratio in the phase transition (see Figure 3(a)), which is accompanied with a rise in the average number of constraint checks. PBMM-EA/DGS starts out over 34 times faster than SAW-EA. When the number of edges increases this difference decreases to 7 times as fast. CLQ-DSATUR seems to have the least problems with this phase transition. However, it shows a sudden rise in constraint checks at 814 edges.

6   Conclusions

We introduced a new model to tackle the graph k-colouring problem that serves as an efficient representation that helps to reduce constraint checks. We verify this efficiency by embedding it in an evolutionary algorithm and performing an empirical comparison on two test suites. The results from the first test suite show a performance in speed and accuracy that is quite favourable, especially on the large real world problem instances with more than 400 nodes. However in the second test, where we look at equipartite graphs in the phase transition, the success ratio shows the typical dip we often observe for stochastic algorithms.

Set against one other popular evolutionary algorithm for constraint satisfaction, the stepwise adaptation of weights method, and against an improved version of a popular complete method, CLQ-DSATUR, the model combined with an evolutionary algorithm that also uses a heuristically guided mutation operator yields promising results. For larger problem instances it is even up to 50 times faster than SAW-EA and CLQ-DSATUR. For difficult equipartite graphs, i.e., that lie in the peak of the phase transition, it is slower and less effective than CLQ-DSATUR, but it is much faster than SAW-EA.

The benefits of the new representation depend highly on the quality of the evolutionary algorithm that is guiding it and the structure of the graph. Further work is needed to tune this evolutionary algorithm and to map the performance against different graph properties.

Acknowledgements. We would like to thank Tezuka, Masaru and Phillip Tann for their collaboration in the first version of the model [19], to Michele Sebag, École Polytechnique and János Csirik, University of Szeged for valuable suggestions, and to Marc Schoenauer for his help with the EO library [12]. This work was supported by the Hungarian National Information Infrastructure Development Program through High Performance Supercomputing as the project CSPAI/1066/2003-2004.

References

1. Tsang, E.: Foundations of Constraint Satisfaction. Academic Press (1993)
2. Johnson, D., Trick, M.: Cliques, Coloring, and Satisfiability. American Mathematical Society, DIMACS (1996)
3. Hertz, A., de Werra, D.: Using tabu search techniques for graph coloring. Computing **39** (1987) 345–351
4. Chams, M., Hertz, A., Werra, D.D.: Some experiments with simulated annealing for coloring graphs. European Journal of Operational Research **32** (1987) 260–266
5. Vesel, A., Zerovnik, J.: How good can ants color graphs? Journal of computing and Information Technology - CIT **8** (2000) 131–136
6. Eiben, A., van der Hauw, J., van Hemert, J.: Graph coloring with adaptive evolutionary algorithms. Journal of Heuristics **4** (1998) 25–46
7. van Hemert, J.: Comparing classical methods for solving binary constraint satisfaction problems with state of the art evolutionary computation. In et al., S.C., ed.: Applications of Evolutionary Computing. Volume 2279 of LNCS. (2002) 81–90
8. van Hemert, J., Bäck, T.: Measuring the searched space to guide efficiency: The principle and evidence on constraint satisfaction. In et al., J.M., ed.: Proceedings of Parallel Problem Solving from Nature VII. Number 2439 in LNCS (2002) 23–32

9. Marino, A., Damper, R.: Breaking the symmetry of the graph colouring problem with genetic algorithms. In Whitley, D., ed.: Late Breaking Papers at the 2000 Genetic and Evolutionary Computation Conference. (2000) 240–245
10. Culberson, J.: Iterated greedy graph coloring and the difficulty landscape. Technical Report TR 92-07, University of Alberta, Dept. of Computing Science (1992)
11. Culberson, J., Gent, I.: Frozen development in graph coloring. Theoretical Computer Science **265** (2001) 227–264
12. Keijzer, M., Merelo, J., Romero, G., Schoenauer, M.: Evolving objects library. Internet Publication (2002)
13. Bäck, T., Fogel, D., Michalewicz, Z., eds.: Handbook of Evolutionary Computation. Institute of Physics Publishing Ltd, Bristol and Oxford University Press, New York (1997)
14. Falkenauer, E.: Genetic Algorithms and Grouping Problems. John Wiley (1998)
15. Trick, M.: Easy code for graph coloring. Internet Publication (1995) Last Modified: November 2, 1994.
16. Brélaz, D.: New methods to color the vertices of a graph. Communications of the ACM **22** (1979) 251–256
17. Mehrotra, A., Trick, M.A.: A column generation approach for graph coloring. INFORMS Journal on Computing **8** (1996) 344–354
18. Trick, M.: Computational series: Graph coloring and its generalizations (2003) http://mat.gsia.cmu.edu/COLORING03.
19. Juhos, I., Tóth, A., Tezuka, M., Tann, P., van Hemert, J.: A new permutation model for solving the graph k-coloring problem. In: Kalmár Workshop on Logic and Computer Science. (2003) 189–199

Hardness Prediction for the University Course Timetabling Problem

Philipp Kostuch[1] and Krzysztof Socha[2]

[1] Oxford University, Department of Statistics
1 South Parks Rd., Oxford, OX1 3TG, UK
kostuch@stats.ox.ac.uk
http://www.stats.ox.ac.uk
[2] IRIDIA, Université Libre de Bruxelles, CP 194/6,
Av. Franklin D. Roosevelt 50, 1050 Brussels, Belgium
ksocha@ulb.ac.be
http://iridia.ulb.ac.be

Abstract. This paper presents an attempt to find a statistical model that predicts the hardness of the University Course Timetabling Problem by analyzing instance properties. The model may later be used for better understanding what makes a particular instance hard. It may also be used for tuning the algorithm actually solving that problem instance. The paper introduces the definition of hardness, explains the statistical approach used for modeling instance hardness, as well as presents results obtained and possible ways of exploiting them.

1 Introduction

Metaheuristics are used nowadays for solving numerous types of optimization problems. However, they always have to be carefully tuned to the particular problem they are to solve. This tuning is often done based on the assumption that problem instances of similar size and similar structure pose similar difficulty to solve them by a given (meta)heuristic algorithm. Hence, the algorithms are run on a set of training instances that resemble the ones that the algorithm is expected to solve later, hoping that it will ensure optimal performance.

Unfortunately, often, especially in case of constrained combinatorial optimization problems, very similar problem instances in terms of size and structure, prove not to be similar in difficulty. One of the problems that exemplifies such behavior is the University Course Timetabling Problem (UCTP).

In this paper we try to define a measure of hardness (i.e. the difficulty of solving) for instances of UCTP. Based on this measure, we notice that very similar instances (i.e. ones created with the same parameters for the instance generator) may vary significantly in their difficulty to be solved by a (meta)heuristic algorithm.

Later, we try to show that in the case of UCTP, the hardness is an intrinsic characteristic of an instance with respect to a given (meta)heuristic algorithm.

J. Gottlieb and G.R. Raidl (Eds.): EvoCOP 2004, LNCS 3004, pp. 135–144, 2004.

We show that based on the analysis of an instance and some statistics, it is possible to predict the hardness of an instance for a given algorithm with reasonable accuracy without actually attempting to solve it. This predicted value may be later used to understand better what makes one instance harder than another, and also to tune the parameters of the algorithm, and hence allow to obtain better results.

The remaining part of this paper is organized as follows. Section 2 describes the UCTP. Section 3 explains the underlying idea and motivation for modeling the instance hardness. Following that, Section 4 introduces the statistical model. In Section 5 the performance of the model is presented and discussed. Eventually, Section 6 discusses the results obtained and indicates possible uses of the hardness prediction model.

2   University Course Timetabling Problem

The problem used to illustrate the thesis of this paper, is the University Course Timetabling Problem (UCTP) [1]. It is a type of constraint satisfaction problem. It consists of a set of n events $E = \{e_1, \ldots, e_n\}$ to be scheduled in a set of 45 timeslots $T = \{t_1, \ldots, t_{45}\}$ (5 days in a week, 9 time-slots a day), and a set of j rooms $R = \{r_1, \ldots, r_j\}$ in which events can take place. Additionally, there is a set of students S who attend the events, and a set of features F satisfied by rooms and required by events. Each student attends a subset of events. A feasible timetable is one in which all events have been assigned a time-slot and a room, so that the following hard constraints are satisfied:

- no student attends more than one event at the same time;
- the room is big enough for all the attending students and posses all the features required by the event;
- only one event is taking place in each room at a given time.

In addition, a feasible candidate timetable is penalized equally for each occurrence of the following soft constraint violations:

- a student has a class in the last slot of the day;
- a student has more than two classes in a row (one penalty for each class above the first two);
- a student has exactly one class during a day.

The infeasible timetables are worthless and are considered equally bad regardless of the actual level of infeasibility. The objective is to minimize the number of soft constraint violations in a feasible timetable. The solution to the UCTP is a mapping of events into particular time-slots and rooms.

The instances we use in our research come from a generator written by Paechter[1].

---

[1] http://www.dcs.napier.ac.uk/~benp

3   Idea and Motivation

The initial motivation of the research presented in this paper comes from submissions to the International Timetabling Competition[2]. While creating metaheuristic algorithms for solving the UCTP instances used in the competition, it became obvious that although the instances are of similar size, the difficulty of solving them varies greatly.

Based on this observation, we developed the hypothesis that the difficulty of solving any given instance of UCTP is a function of some parameters of the instance that are intrinsic to it, and more complex than just its size [2]. If this hypothesis is true, it should be possible to define a hardness of an instance in terms of some of its characteristics. It should also be possible to predict values obtained by an algorithm just by looking at the instance.

We should make more precise here, what we mean exactly by the hardness of an instance. In principle, all the instances of UCTP under investigation are known to have at least one perfect solution. Hence, if we defined the hardness as the measure of how close to 0 any instance can be solved, the hardness of all those instances would be exactly the same. In our case however, the goal was to solve the given instances as well as possible, but within a given time limit. We will hence define the hardness of the instance as the quality of the solution obtained within this time limit by a given algorithm. Also, as the results obtained by a (meta)heuristic algorithm for a given problem instance tend to have some variance, we will consider in our investigation only the mean of the obtained values as the measure of hardness.

Following the hypothesis that the hardness depends on some intrinsic characteristic of an instance, we decide to measure several different features of the instances–the model's explanatory variables or covariates. If we find the right covariates, and if the hypothesis is true, the combination of covariates may be used to create a regression model for instance hardness–that is, find the functional relationship between the combination of covariates and the hardness of the instance.

Considering the problem investigated, we chose to measure some summary statistics of the instance to be used as covariates. Altogether there are 27 distinct covariates that we initially consider as important for the regression model. Sec. 4 describes in more detail how the final subset of those has been chosen for the final model.

For the purpose of choosing and fitting the model we used the results obtained by $\mathcal{MAX}$-$\mathcal{MIN}$ Ant System ($\mathcal{MMAS}$) [3] designed particularly for this problem [4]. $\mathcal{MMAS}$ is a version of Ant Colony Optimization (ACO) metaheuristic initially proposed by Dorigo [5]. ACO has been in recent years widely used for solving combinatorial optimization problems [6] also including constrained satisfaction problems [7].

The final model chosen will obviously apply only to the results obtained by our $\mathcal{MMAS}$. Certainly, a comparison of models found for different algorithms

[2] http://www.idsia.ch/Files/ttcomp2002/

may provide some information whether the same features of the instance make it difficult for different algorithms. However interesting this topic might be, we do not focus on it in this paper.

4   Statistical Model

4.1   Normal Linear Models

The task formulated in the previous section is the archetypal statistical problem of regression. Among the large variety of models developed in the regression context, the best known are linear models [8]. The formulaic description for the i-th observation under a linear model is

$$y_i = \sum_{j=1}^{p} \beta_j x_{i,j} + \epsilon_i \tag{1}$$

where y_i is the observed response, $x_{i,j}$'s the covariate values for the i-th observation, β_j's the model parameters and ϵ_i the error of the i-th observation.

The term linear refers to the fact that the parameters appear only linearly in the predictor $\eta(x, \beta) = \sum_{j=1}^{p} \beta_j x_{i,j}$. This does however not mean that such a model can only cater for linear influences of covariates. It is for example possible to incorporate a covariate x in a linear and a quadratic term x^2 if this seems to be justified based on domain specific knowledge. Also possible are so-called interaction terms, i.e. multiplicative effects between covariates. All a linear model requires is for these terms to have a linear parameter. It is also common that in a linear model the predictor acts on the same scale as the observation, by which we mean that the function linking the predictor to the observation is the identity (as has already been assumed in Eq. 1).

To complete a regression model, a distribution for the error terms has to be specified. In normal linear models–also just called normal models–the ϵ_i are assumed to be independent, identically distributed (i.i.d.) according to $N(0, \sigma^2)$.

4.2   Casting Our Problem into a Normal Model Framework

To apply the normal model approach to a given problem, one has to ensure that the underlying modeling assumptions are met. We will do so in this section for the prediction of hardness based on instance characteristics.

The first observation for our data is that we are confronted with a more complex problem than in Eq. 1. Namely, we can separate the error into two terms. One being the standard modeling error associated with a given realization of the covariates. The other being a measurement error of the response as different metaheuristic runs will–on the same instance–yield different results. This can be formulated as

$$y_{i,k} = \sum_{j=1}^{p} \beta_j x_{i,j} + \epsilon_i + \tilde{\epsilon}_{i,k} \tag{2}$$

where $y_{i,k}$ is the k-th measurement on the i-th instance, $x_{i,j}$'s the covariate values for the i-th observation, β_j's the model parameters, ϵ_i the error of the i-th instance and $\widetilde{\epsilon}_{i,k}$ the measurement error for the k-th measurement on the i-th instance.

This kind of model is known as a linear mixed effects model in the literature [9]. Instead of following this approach, we circumvent the problem of two different error terms by averaging out the measurement error $\widetilde{\epsilon}_{i,k}$. For this, we have 15 runs on every instance and use the mean value as response. This allows us to stay within the normal model framework. It means a loss of efficiency in the modeling process as we throw away information that could otherwise be utilized, but eliminates the measurement error that in the algorithm under study is fairly large (see next paragraph). So, the model for the mean values has smaller variation in the error term which opens up the chance of better predictions.

To assess whether averaging over 15 instances is enough to neglect the measurement error in the further treatment of the data we have 50 runs on a particular instance and calculate the means for 25 random samples of size 15. The mean for all 50 values was 150, while the upper and lower quantiles of the means based on 15 runs were 145 and 155 respectively with 3 outliers being further away. To put the mean values in relation to the individual measurements and to give an idea of the measurement error itself, we note the lower and upper quantiles for the 50 runs to be 135 and 160.

So, while there is still some variation, this spread is small enough to continue with our approach, noting that the remaining measurement error will be a lower bound on the achievable prediction quality. This particular instance was chosen as the algorithm could only achieve relatively high values on it and a consistent observation over all instances was that the variance of the measurement error increases with the response value.

This heteroscedasticy of the measurement errors is the inspiration for the second concern we have fitting the data with a normal model: the distribution assumption for the remaining error term. As mentioned in Sec. 4.1 the ϵ_i's are assumed to be i.i.d. $\sim N(0, \sigma^2)$. Two things can go wrong with this. The errors may not be normally distributed or they may not have constant variance. In particular, the error variance may be a function of the response value as was observed for the measurement error, where it grew with the response. Unfortunately, with the ϵ_i's we are not in a position to test this a priori. It will only be possible to assess the appropriateness of the distribution assumption a posteriori via diagnostic tools on the residuals of the fitted model. We will do this in Sec. 5.

We mention two techniques that deal with such deviations from the normal model. The first is to achieve constant variance by transforming the response to a different scale, e.g. using the log function on it before fitting the model [10]. As we will implement this, we shall explain the rational behind such a transformation. If we assume the error not to be an additive term, but rather a fraction of the response, taking the log will transform a multiplicative error into an additive one that fits into the normal model framework. If the problem lies deeper, i.e. a non-normal error distribution, Generalized Linear models (for example with a Poisson distribution) have to be used [11].

4.3   Fitting the Model

This section deals with the non-mathematical side of model fitting. By mathematical model fitting we mean the process of determining the parameter values β when all other properties of the model are fully specified. The mathematics behind this process are simple as for normal models minimizing the log-likelihood (which is the fitting approach for more general models) coincides with minimizing the sum of squares (SSQ) of the errors.

What we concern ourselves with, is the process of selecting a subset of the available covariates to form the final model [12]. The need for covariate selection arises from the purpose for which we model: prediction. We have a training set of 120 new instances generated from the range of input variables to the instance generator that were used for the competition instances. If we fit the model with all covariates, we get a lower bound on how good we can fit this set, but the prediction quality will be poor as we have over-fitted our data, see Table 1. Therefore, the following procedure has been implemented in S-Plus for both the model with unchanged response and the one with the log-response.

Simulated Annealing [13] is used to determine a good set of covariates. This is done as the search space–especially when interactions are included–gets too large for enumeration. Standard covariate selection methods such as forward and backward selection are greedy methods that can get stuck in bad local optima. As search space we have all possible subsets of the full covariate set. We choose a random starting location, and the neighborhood is defined by either randomly dropping or adding a covariate. Because of the expensive fitness evaluation, we can only run a moderate number of iterations.

For the fitness evaluation of a solution, 6-fold cross-validation is used on the training set, i.e. it is randomly divided in 6 sets of 20 instances, and then 5 sets are used to fit the current model and the remaining set is used to assess the prediction power by recording the error SSQ for this 6th set. This is repeated for leaving out all sets in turn. The fitness for the current solution is the average over the 6 values. One modification is put in place when the predicted value for an instance is negative: negative predictions are set to 0 before we calculate the error term. This is reasonable as a negative value in the prediction can be interpreted as an instance that is solved to optimality in less time than allowed.

The Simulated Annealing engine is run 25 times and for every run we record the covariates included in the final model. Then we drop those covariates that were included only 3 or less times in these final models. On the remaining set of covariates we restart the whole procedure leading after another 25 runs to a second set of covariates that are to be dropped for good. In both models this turns out to be sufficient to end up with a stable set of covariates, as Simulated Annealing on the remaining covariates did not drop anymore.

5   Results

With the above approach we arrive at two final models: one for the unchanged response and the other for the log response. Two reasons let us focus on the log model. Firstly, analytical plots for the residuals indicate that the log model meets the normality assumption better. Secondly, we observe that the fitting criterion is different for the two models, with the log model pursuing the–in our opinion–more reasonable goal.

We explain this difference briefly. Both models were fitted such that they minimize a fitness function based on the error SSQ, i.e. on the gap in absolute values between prediction and observation. But since the modeled values are on different scales, this leads to different fitting goals on the original response scale. The model with unchanged response controls the spread around the observation in terms of absolute value. In contrast, the error for the log response enters the model on the original scale as $prediction \times e^{\epsilon_i}$, where for small errors $e^{\epsilon_i} \approx 1 + \epsilon_i$ holds. So, the log response model minimizes approximately the squared sum of percentage errors.

Another observation of interest in this context is that using squared error values leads to a worse average error in favor of limiting extreme errors. In the case of a normal distribution, which we assume for our errors, the so calculated standard deviation σ has a ready interpretation as signifying a confidence level of approximately 70%.

Table 1 summarizes the results of the models fitted on the log response. The first two are the models with all covariates included (Full) and the model reduced by Simulated Annealing (Final). The last model (Interaction) will be introduced later. In the final model we dropped 1 more term after the automated selection, as it was statistically insignificant. As a test set, we used the 20 competition instances.

We see that the full model fits the training set better but its prediction power is worse. For both models the prediction error is larger than the error on the training set, and clearly larger than the predicted error $\hat{\sigma}$ based on the cross-validation. We note that this prediction is biased and will always under-estimate the real error. The average percentage errors have been included as they are more intuitively understandable, but we point out that optimization has been carried out with respect to σ-values.

Table 1. Summary of the errors in the 3 models fitted on log(response). Values are the %-errors between prediction and observation on the original scale.

Model	parameters	Training set		Test set		
		σ	Av. Error	$\hat{\sigma}$	σ	Av. Error
Full	27	18.0	13.2	25.7	28.4	21.8
Final	9	19.4	13.8	21.0	27.2	19.8
Interaction	18	17.2	12.4	21.0	22.0	16.6

We will now discuss the final model and its remaining covariates in more detail. Table 2 gives the 8 covariates for this model, the covariate parameters and their values on the the response scale, i.e. e^{β_i}. The covariate values–except for slack–have been normalized, i.e. were transformed to have 0 mean and variance

1. This allows a direct comparison of the effect size independent of the scaling of the covariates. Slack was not subjected to this procedure as it is a discrete covariate with only two levels.

Table 2. Final set of 8 covariates used by the regression model ordered by their level of importance.

Name	Coefficient	Effect on Response scale
av. weighted event degree	1.39	4.02
av. event size	-1.15	0.32
slack	-1.22	0.29
av. weighted room options	-0.63	0.53
sd. events per room	0.62	1.85
av. event degree1	0.51	1.66
sd. event degree2	0.27	1.31
no of 1-option events	0.20	1.22

Before we can interpret the covariates we have to introduce a graph closely related to the UCTP. Events are the nodes in this graph and they are joined by an edge if the events cannot be placed in the same time-slot. There can be two reason for an edge: a student who attends both events or the two events need both the same room. The original graph (referred to as 1) comprises only student conflicts while the second graph (referred to as 2) includes room conflicts as well.

Four effects in the model are related to the node degree in this graph. The weighted event degree (the weights are the number of students that cause a given edge) and the ordinary event degree in graph 1 increase the prediction as they grow. The effect of the weighted event degree seems to be larger but in fact it is almost perfectly correlated with the average event size. This means that in a normal instance these two effects will have to be subtracted giving a much smaller effect for the average weighted event degree. Dropping either one of the correlated terms and hoping that the other would absorb the effect of both was considered but led to a worse fit. The last effect referring to this graph is the standard deviation in event degree in graph 2. The problem becomes harder as the spread increases.

Three effects are related to the room constraints. Most prominently, the average weighted number of room options per event (the square root was used for weighting to put emphasis on small numbers) is negatively correlated with the hardness. Positively correlated is the number of events that have only one room options. We recorded them as they form a special subset of events, causing additional edges between graph 1 and graph 2. The third covariate measures the standard deviation in the number of possible candidate events from the room's point of view. The more spread out this distribution is the harder the problem.

The last effect in the model-and the most decisive-is the factor variable slack, which indicates whether 40 time-slots are a lower bound for the placement of the events, i.e. whether the number of rooms times 40 equals the number of events. If slack is present, we can expect the problem to be substantially easier. For this statement-as for all others made in this section-we assume a *ceteris paribus* situation, i.e. that we can change 1 covariate at a time keeping the others fixed. Obviously, this assumption is hard to meet in a real problem and therefore this effect analysis has to be seen with this limitation in mind.

After isolating these 8 covariates, we included them and all 2nd order interactions in another normal model. This model was then pruned according to a standard statistical criterion (AIC stepwise selection, taken from the MASS library [10], which is a greedy search method that trades off the log likelihood of a model against the number of covariates used). We note that this criterion optimizes prediction power only indirectly. The results for this model can be seen in Table 1. It achieves a better fit on the training set than the full model and a better prediction than the final model. The gap between the predicted error and the observed error has markedly decreased. This clearly indicates that further research should be directed towards more complicated models.

6   Conclusions

The paper presents a successful attempt to predict the quality of the solutions found by the $\mathcal{MAX}$-$\mathcal{MIN}$ Ant System for the University Course Timetabling Problem. We have attempted to predict the hardness of a complex constrained optimization problem for a metaheuristic algorithm. Without making any explicit assumptions about the inner operation of the algorithm (i.e. we treated it as a black box algorithm), we have managed to develop a reasonably simple regression model allowing to predict the competition results with an average error of less than 17%.

The statistical model we developed may be further exploited in at least two distinctive ways. One option is to use the knowledge acquired in order to modify the algorithm and improve its performance. This may include exploiting the understanding of important covariates, or exploiting the estimated hardness. The second option is to use the model for sensitivity analysis of the instance. When considering the UCTP as a real world problem, it may be possible to change some characteristics. Based on the model, the changes with the best marginal benefit can be determined (e.g. should one more room be made available for the courses or is it better to fit the existing rooms with additional features).

6.1   Future Work

In the future we plan to investigate alternative model choices that we left for now unexplored. This includes linear models with a more systematic interaction investigation, generalized linear models and non-linear models.

Independently, we would like to investigate in more detail, how the predicted hardness may be used for improving the performance of the algorithm. Also, we would like to develop similar models for other algorithms solving the UCTP. This way we could see if it is possible to maintain this level of prediction accuracy for the other algorithms, and also if the covariates that are important for one algorithm are similarly important for the others. This research will show whether the concept of hardness can be to some extend generalized, i.e. can be made less algorithm-dependent.

Acknowledgments. Philipp Kostuch acknowledges the support of a Marie Curie Training Site fellowship funded by the Improving Human Potential (IHP) programme of the Commission of the European Community (CEC), grant HPRN-CT-2000-00032. The research work was also supported by the "Metaheuristics Network", a Marie Curie Research Training Network funded by the Improving Human Potential programme of the CEC, grant HPRN-CT-1999-00106, and by the "ANTS" project, an "Action de Recherche Concertée" funded by Scientific Research Directorate of the French Community of Belgium.

References

1. Rossi-Doria, O., Sampels, M., Chiarandini, M., Knowles, J., Manfrin, M., Mastrolilli, M., Paquete, L., Paechter, B.: A comparison of the performance of different metaheuristics on the timetabling problem. In Burke, E.K., De Causmaecker, P., eds.: Proceedings of PATAT 2002. Volume 2740. (2003)
2. Kostuch, P.: University Course Timetabling, Transfer Thesis, Oxford University, England (2003)
3. Stützle, T., Hoos, H.H.: $\mathcal{MAX}\text{-}\mathcal{MIN}$ Ant System. Future Generation Computer Systems **16** (2000) 889–914
4. Socha, K.: $\mathcal{MAX}\text{-}\mathcal{MIN}$ Ant System for International Timetabling Competition. Technical Report TR/IRIDIA/2003-30, Université Libre de Bruxelles, Belgium (2003)
5. Dorigo, M., Di Caro, G.: The Ant Colony Optimization meta-heuristic. In Corne, D., Dorigo, M., Glover, F., eds.: New Ideas in Optimization, McGraw-Hill (1999)
6. Maniezzo, V., Carbonaro, A.: Ant Colony Optimization: an Overview. In Ribeiro, C., ed.: Essays and Surveys in Metaheuristics, Kluwer Academic Publishers (2001)
7. Roli, A., Blum, C., Dorigo, M.: ACO for maximal constraint satisfaction problems. In: Proceedings of the 4th Metaheuristics International Conference (MIC 2001). Volume 1. (2001) 187–191
8. Neter, J., Wasserman, W., Kutner, M.: Applied Linear Statistical Models. 3 edn. Irwin (1990)
9. Davidian, M., Giltinan, D.: Non-linear Models for Repeated Measurement Data. Chapman & Hall (1995)
10. Venables, W., Ripley, B.: Modern Applied Statists with S-PLUS. 3 edn. Springer (1999)
11. McCullagh, P., Nelder, J.: Generalized Linear Models. 2 edn. Chapman & Hall (1989)
12. Hastie, T., Tibshirani, R., Friedman, J.: The Elements of Statistical Learning. Springer (2001)
13. Kirkpatrick, S., Gelatt, C.D., Vecchi, M.P.: Optimization by Simulated Annealing. Science **220** (1983) 671–680

Hybrid Estimation of Distribution Algorithm for Multiobjective Knapsack Problem

Hui Li, Qingfu Zhang, Edward Tsang, and John A. Ford

Department of Computer Science, University of Essex
Wivenhoe Park, Colchester CO4 3SQ, United Kingdom
{hlil,qzhang,edward,fordj}@essex.ac.uk

Abstract. We propose a hybrid estimation of distribution algorithm (MOHEDA) for solving the multiobjective 0/1 knapsack problem (MOKP). Local search based on weighted sum method is proposed, and random repair method (RRM) is used to handle the constraints. Moreover, for the purpose of diversity preservation, a new and fast clustering method, called stochastic clustering method (SCM), is also introduced for mixture-based modelling. The experimental results indicate that MOHEDA outperforms several other state-of-the-art algorithms.

1 Introduction

Over the past twenty years, numerous multiobjective evolutionary algorithms (MOEAs) have been proposed for multiobjective optimization problems (MOPs) [2]. Compared with classical methods, MOEAs are more suitable for solving MOPs for the following reasons: (i) multiple solutions can be found in a single run of a MOEA; (ii) a good spread of the nondominated solutions can be reached; and (iii) a MOEA is less susceptible to the shape or continuity of the Pareto-optimal front. Due to the conflicting relationships between objectives, it is unlikely to find such a solution that optimizes all objectives simultaneously. In practice, it is a hard task to find all nondominated solutions when, as is often the case, the number of Pareto-optimal solutions is huge or even infinite. Therefore, when applying an evolutionary algorithm to MOPs, two major issues should be addressed:

- The resultant nondominated front should be as close to the Pareto-optimal front as possible (convergence);
- The nondominated solutions should be distributed as uniformly (in most cases) and widely along the Pareto-optimal front as possible (diversity).

The performance of MOEAs can be improved by local search. IMMOGLS [5] should be regarded as the first algorithm to hybridize MOEAs with local search. In this algorithm, all weights are generated randomly. Jaszkiewicz (1998) proposed multiobjective genetic local search (MOGLS)[1]. Weighted Tchebycheff scalarizing functions are used to evaluate the fitness value for each individual solution. Local search and selection are implemented on the basis of the current

J. Gottlieb and G.R. Raidl (Eds.): EvoCOP 2004, LNCS 3004, pp. 145–154, 2004.
© Springer-Verlag Berlin Heidelberg 2004

scalarizing function. In MOGLS, all weights of the scalarizing function are also created in a random way. Without using the linear combination of objectives, Knowles and Corne (2000)[7] proposed a memetic Pareto archived evolutionary strategy (M-PAES), which combines the recombination operators with an evolutionary strategy (PAES). In M-PAES, the acceptance of a new solution generated by genetic search or local search depends not only on the Pareto-dominance relationship but also on the grid-type partition of the objective space.

Estimation of distribution algorithm (EDA) is a new paradigm in evolutionary computation. Recently, few multiobjective EDAs have been proposed and studied. Mixture-Based Iterated Density Estimation Evolutionary Algorithms (MIDEA) was proposed by Thierens and Bosman (2001)[3]. Multiobjective Bayesian Optimization Algorithms (mBOA) was introduced by Khan, Goldberg, and Pelikan (2002)[9]. Population-Based Ant Colony Optimization for MOP was developed by Guntsch1 and Middendorf (2003)[8]. This class of algorithms generalizes genetic algorithms by replacing the crossover and mutation operators with learning and sampling the probability distribution of the best individuals in the previous population.

However, the main goal of MOEAs is to improve the performance both in algorithmic convergence and in diversity preservation. To this end, we propose a hybrid estimation of distribution algorithm for solving the MOKP.

This paper is organized as follows. In section 2, MOPs and MOKP are briefly introduced. Local search is discussed in section 3. Section 4 contains stochastic clustering method. In the following section, MOHEDA is proposed. The experimental results are presented in section 6. In the final section, conclusions are given.

2   Multiobjective Optimization

A general MOP consists of a set of d decision variables, a set of n objectives, and a set of m constraints. Both objectives and constraints are functions of the decision variables. The optimization goal is to

$$
\begin{aligned}
\text{maximize} \quad & f_i(\mathbf{x}) & i = 1, \dots, n \\
\text{s.t.} \quad & c_j(\mathbf{x}) \leq 0 & j = 1, \dots, m
\end{aligned} \tag{1}
$$

where $\mathbf{x} = (x_1, \dots, x_d)$ is the decision vector in the decision space and $\mathbf{f}(\mathbf{x}) = (f_1(\mathbf{x}), \dots, f_n(\mathbf{x}))$ is the objective vector in the objective space.

For any two decision vectors $\mathbf{x}^{(1)}$ and $\mathbf{x}^{(2)}$, $\mathbf{x}^{(1)}$ is said to dominate $\mathbf{x}^{(2)}$, denoted by $\mathbf{x}^{(1)} \succ \mathbf{x}^{(2)}$, if $f_k(\mathbf{x}^{(1)}) \geq f_k(\mathbf{x}^{(2)}), \forall k \in \{1, \dots, n\}$ and there exists at least a $k' \in \{1, \cdots, n\}$ such that $f_{k'}(\mathbf{x}^{(1)}) > f_{k'}(\mathbf{x}^{(2)})$. A decisoin vector x^* is said to be Pareto-optimal if there is no any decision vector in the decision space which dominates x^*.

MOKP is a well-known NP-hard multiobjective combinatorial optimization problem, which has been studied by many researchers for testing the performance of MOEAs. It is a generalization of the 0-1 single objective knapsack problem.

Let p_{ij} be the profit of item j associated with knapsack i, w_{ij} be the weight of item j associated with knapsack i, and c_i be the capacity of knapsack i, the goal is to find a binary vector $\mathbf{x} = (x_1, \ldots, x_d) \in \{0, 1\}^d$ such that

$$\sum_{j=1}^{d} w_{ij} \cdot x_j \le c_i, i = 1, 2, \ldots, m \qquad (2)$$

and for which $\mathbf{f}(\mathbf{x}) = (f_1(\mathbf{x}), \ldots, f_n(\mathbf{x}))$ is maximized, where

$$f_i(\mathbf{x}) = \sum_{j=1}^{d} p_{ij} \cdot x_j, i = 1, \cdots, n. \qquad (3)$$

3   Local Search

Local search is a simple iterative method for finding good approximate solutions. It should be pointed out that the acceptance function of a neighbour solution in MOPs are slightly different from that in single objective problems. For example, in M-PAES, a reference set is required to evaluate the Pareto-based fitness values of solutions. In this paper, the acceptance function $h(\mathbf{x})$ is defined as the linear combination of all objectives by:

$$h(\mathbf{x}) = \sum_{i=1}^{n} w_i \cdot f_i(\mathbf{x}) \qquad (4)$$

where $w_i > 0, i = 1, \ldots, n$, is the weight for the ith objective and $\sum_{i=1}^{n} w_i = 1$.

In IMMOGLS and MOGLS, the weights for the acceptance function are generated randomly. Here, the weights are determined by the start solution $\mathbf{x}$ in the following way. Two extra vectors in the objective space: $(f_{max}^{(1)}, \ldots, f_{max}^{(n)})$ and $(f_{min}^{(1)}, \ldots, f_{min}^{(n)})$, are required. In the context of evolutionary algorithm, $f_{max}^{(k)}$ and $f_{min}^{(k)}$ are the maximization and the minimization of the kth objective in the population. Then, the raw weight w_i' is given by:

$$w_i' = f_i(\mathbf{x}) - f_{min}^{(i)}, i = 1, \ldots, n. \qquad (5)$$

Take the order of magnitude into consideration, the unbiased coefficient $e_i, i = 1, \ldots, n$, is introduced as:

$$e_i = \frac{1}{f_{max}^{(i)} - f_{min}^{(i)}}. \qquad (6)$$

Therefore, the weights for the acceptance function are:

$$w_i = \frac{w_i' \cdot e_i}{\sum_{j=1}^{n} w_j' \cdot e_j}, i = 1, \ldots, n. \qquad (7)$$

The local search procedure for MOKP is described as follows:

Algorithm 1: Local Search

- (1) Select the starting solution $\mathbf{x}$, and generate a set of heuristic weights $w_i, i \in \{1, \dots, n\}$ according to (7);
- (2) For each item $j \in I = \{n_1, \dots, n_{|I|}\}$ with $x_j = 0$;
  - Set $x_j := 1$;
  - Repair the modified $\mathbf{x}$;
  - Record the set $\mathcal{I}_j$ of all removed items;
  - Compute $t_j := \sum_{i=1}^{n} w_i \cdot p_{ij} - \sum_{k \in \mathcal{I}_j} \sum_{i=1}^{n} w_i \cdot p_{ik}$;
  - Reset $\mathbf{x}$.
- (3) Sort the series $t_j, j \in \{n_1, \dots, n_{|I|}\}$ and calculate $j_m := arg \max_{j \in I} t_j$;
- (4) Update $\mathbf{x}$ with $x_{j_m} := 1$ and set $x_k := 0$ for all $k \in \mathcal{I}_{j_m}$;
- (5) If the maximal number of local search steps is reached, stop; Otherwise, goto step 2.

Note that a portion of solutions generated by recombination operators or the move in local search might be infeasible. In this case, some repair strategies should be employed to handle the constraints. Greedy repair method (GRM) is a frequently-used approach. In this paper, we propose random repair method (RRM) for handling constraints and constructing the random neighbourhood. The items will be randomly removed from the knapsacks until the constraints are satisfied. In comparison with GRM, RRM is easy to implement and its computational time is less than that of GRM. We also note that the initial population with higher quality can be created by GRM.

4   Stochastic Clustering Method

As pointed out in [3], the diversity of the Pareto-optimal front can be maintained by the mixture of factorized probability distributions. The population is divided into a couple of clusters using the Euclidean leader algorithm. Each cluster is processed separately and a local probability model is built on its related cluster. In the context of population-based algorithms, each cluster consists of a number of good selected solutions which are currently optimal with respect to a certain weighted sum function. Then, different possible weighted sum functions can be simultaneously optimized in a single generation. In this paper, we propose a new and fast clustering method, called stochastic clustering method (SCM). The process of SCM is mathematically described in the remainder of this section.

Assume that $\prod_{i=1}^{n}[LB_i, UB_i]$ is the domain of a population P in the objective space, where $LB_i = \min_{\mathbf{x} \in P} f_i(\mathbf{x})$ and $UB_i = \max_{\mathbf{x} \in P} f_i(\mathbf{x})$. Suppose that we have a set of subdomains $D_j = \prod_{i=1}^{n}[LB_i^{(j)}, UB_i^{(j)}], j = 1, \dots, J$, where J is the number of subdomains. A quantitative metric for the objective with the maximal range in the subdomain $D_j, j = 1, \dots, J$, is defined as:

$$r_j = \max_{i \in \{1, \dots, n\}} \frac{UB_i^{(j)} - LB_i^{(j)}}{UB_i - LB_i}. \tag{8}$$

The index of the objective with the maximal range in the subdomain D_j is given by:

$$I_j = arg \max_{i \in \{1,\dots,n\}} \frac{UB_i^{(j)} - LB_i^{(j)}}{UB_i - LB_i}. \tag{9}$$

Then, the Kth ($K = arg\max_{j \in \{1,\dots,J\}} r_j$) subdomain with the maximal metric for the range is decomposed in the following way:

$$
\begin{aligned}
LB_i^{(a)} &= LB_i^{(b)} = LB_i^{(K)} \text{ if } i \neq I_K, \\
UB_i^{(a)} &= UB_i^{(b)} = UB_i^{(K)} \text{ if } i \neq I_K, \\
LB_{I_K}^{(a)} &= LB_{I_K}^{(K)}, \\
UB_{I_K}^{(a)} &= LB_{I_K}^{(K)} + (\frac{1}{3} + \alpha)\Delta, \\
LB_{I_K}^{(b)} &= LB_{I_K}^{(K)} + (\frac{1}{3} + \alpha)\Delta, \\
UB_{I_K}^{(b)} &= UB_{I_K}^{(K)}
\end{aligned}
\tag{10}
$$

where $\Delta = UB_{I_K}^{(K)} - LB_{I_K}^{(K)}$ and α is a random number in $[0, 1]$. In this way, the subdomain D_K can be divided into two nonoverlapped subdomains $D_a = \prod_{i=1}^{n}[LB_i^{(a)}, UB_i^{(a)}]$ and $D_b = \prod_{i=1}^{n}[LB_i^{(b)}, UB_i^{(b)}]$. In SCM, there is no need to calculate the distances between individuals which are crucial for other clustering algorithms. Moreover, subdomains dynamically vary with generations.

For instance, we have two two-dimensional subdomains $D_1 = [0, 10] \times [0, 20]$ and $D_2 = [0, 20] \times [0, 30]$. Here, we assume that e_1 equals to e_2. Then, the metric values of D_1 and D_2 are $20 \cdot e_1$ and $30 \cdot e_2$ respectively. Therefore, we decompose D_2 into $[0, 20] \times [10 + \alpha 10, 30]$ and $[0, 20] \times [0, 10 + \alpha 10]$.

5   Hybrid Estimation of Distribution Algorithm

Estimation of Distribution Algorithms (EDAs) were first introduced by Mühlenbein and Paaß (1996) [6]. The detailed introduction of EDAs can be found in Larrañaga's book [10]. In EDAs, there are neither crossover nor mutation operators. Instead, a new population is sampled from a probability distribution, which is modelled from the selected best solutions. The dependencies among the variables involved can be explicitly and effectively captured and exploited. According to the types of dependencies among variables involved, EDAs can be classified into three categories: without dependencies (UMDA 1998, PBIL 1994, and cGA 1998), bivariate dependencies (MIMIC 1997, COMIT 1997, and BMDA 1999), and multivariate dependencies (EcGA 1999, FDA 1999, EBNA 2000, and BOA 2000).

Recently, hybrid estimation of distribution algorithms have been developed for solving global optimization [11]. To improve the performance of a MOEA both in convergence and in diversity, we propose a mixture-based hybrid estimation of distribution algorithm (MOHEDA). Univariate Marginal Distribution Algorithm (UMDA) is employed. In UMDA, the joint probability distribution of the selected individuals at each generation, $p(\mathbf{x}, t)$, is factorized as a product of independent univariate marginal distributions $p(x_i, t), i = 1, \ldots, d$. This joint distribution can be written as:

$$p(\mathbf{x}, t) = p(\mathbf{x}|Pop^{Se}(t-1)) = \prod_{i=1}^{d} p(x_i, t) \qquad (11)$$

where $Pop^{Se}(t-1)$ is the set of selected individuals from the previous population.

In MOHEDA, an elite population is maintained to store all nondominated solutions found so far. At the beginning of search, this population is empty. It is updated by the local optima obtained in the local search. The mating pool $P_m^{(t)}$ consists of the solutions selected from the elite population or current population. Using SCM, the domain of $P_m^{(t)}$ is divided into n_d subdomains $D_j, j = 1, \ldots, n_d$. The mixture-based probability models can be reformulated by:

$$p^{(k)}(\mathbf{x}, t) = p(\mathbf{x}|Pop^{Se}(k, t)) = \prod_{i=1}^{d} p^{(k)}(x_i, t) \qquad (12)$$

where $Pop^{Se}(k, t)$ is the set of mating parents which belong to the kth subdomain D_k. In the following, we summarize the framework of MOHEDA.

Algorithm 2: MOHEDA

- (1) Initialization: Generate an initial population $P^{(0)}$ with n_p individuals randomly and create an elite population $P_e^{(0)} := \emptyset$; Perform local search on $P^{(0)}$ and update $P_e^{(0)}$; Set t: =0.
- (2) Selection: Select n_p individuals from $P^{(t)} \cup P_e^{(t)}$ and place them in the mating pool $P_m^{(t)}$;
- (3) Stochastic Clustering: Decompose the domain of $P_m^{(t)}$ into n_d disjoint subdomains;
- (4) Mixture-Based Modelling: Estimate the distribution $p^{(k)}(\mathbf{x}, t)$ on each subdomain $D_k, k = 1, \ldots, n_d$;
- (5) Sampling: Sample the new population $P^{(t+1)}$ with n_p individuals according to the distributions $p^{(k)}(\mathbf{x}, t), k = 1, \ldots, n_d$;
- (6) Local Search: Select $\frac{n_p}{2}$ individuals from $P^{(t+1)}$, perform local search for each selected individual and update $P_e^{(t+1)}$ with the local optima;
- (7) Termination: If the termination criteria are satisfied, then stop. Otherwise, set t: =t+1 and go to step 2.

In MOHEDA, both repair methods are taken into account. In the initialization, we use GRM to generate high quality solutions. Besides, new solutions produced by sampling and local search are repaired by RRM. Note that the size of the elite population will probably be increased since more and more new possible nondominated solutions may be found during the evolution. To maintain the limited size n_e of the elite population, we remove some individuals randomly until the number of elite individuals is less than n_e.

6   Experiments

In this section, the performance metrics of MOEAs are introduced. Some experiments are devised to test the performance of MOHEDA both in convergence and in diversity. The numerical results are briefly discussed.

6.1   Performance Metrics

How to assess the performance of MOEAs is very difficult. Many performance metrics have been proposed and studied for convergence and diversity. In this section, three metrics are briefly discussed.

The coverage metric proposed by Zitzler is used to assess the performance of various MOEAs. The function $\mathcal{C}(A, B)$ maps the ordered algorithm pair (A, B) into a real number in $[0, 1]$. The value $\mathcal{C}(A, B)$ means the percentage of the solutions in A covered by the solutions in B.

In order to evaluate the convergence of MOGLS and MOHEDA, the average values (front error) of the minimal, mean and maximal distances between nondominated front and the Pareto-optimal front in 20 runs are calculated. The lower value of the distances, the closer to the Pareto-optimal front the resultant nondominated front is.

Consider not only the standard deviation of distances between nondominated solutions but also the maximal spread, we propose a new metric to evaluate the spread of the set A of nondominated solutions. This metric can be formulated as:

$$D = \frac{\sum_{k=1}^{n}(f_{max}^{(k)} - f_{min}^{(k)})}{\sqrt{\frac{1}{|A|}\sum_{i=1}^{|A|}(d_i - \bar{d})^2}} \tag{13}$$

where d_i is the Euclidean distance between the ith solution and its closest neighbour and $\bar{d}$ is the mean distance. The greater value of the diversity metric, the better the diversity of nondominated solutions is.

6.2   Experimental Settings

The proposed MOHEDA has been compared with MOGLS. Nine test instances including 2, 3 and 4 objectives, associated with 250, 500 and 750 items respectively, are considered. The profits, weights and capacities in our experiments are identical to the data reported in [1][3], which can be downloaded from the link http://www.tik.ee.ethz.ch/~zitzler/testdata.html/.

In Jaszkiewicz's paper [1], MOGLS has been demonstrated to be the best algorithm for MOKP. MOGLS outperforms several other MOEAs, such as NSGA, SPEA and M-PAES. Thus, we only compared our algorithm with MOGLS. MOHEDA was implemented in C++. All experiments were performed on identical PCs (AMD Athlon 2400MHZ) running Linux.

Three parameters: the size of the population n_p, the number of subdomains n_d and the limited size of the elite population n_e, are given in Table 1. To be

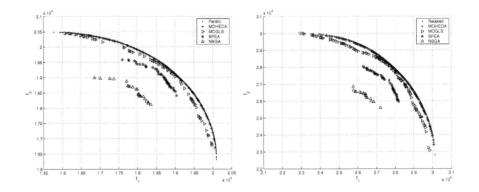

Fig. 1. Nondominated solutions of the instances with 2 knapsacks: 500 items (left), 750 items (right).

consistent with the parameter settings of other algorithms , the size of population and the maximal number of objective function evaluations are determined in the same way as SPEA [4] and MOGLS [1]. The search was terminated after completing $n_p \times 500$ objective function evaluations. We performed MOHEDA 20 runs on each of 9 instances.

Table 1. Parameter Settings

Items	250			500			750		
Parameters	n_p	n_d	n_e	n_p	n_d	n_e	n_p	n_d	n_e
2 Knapsacks	150	15	500	200	20	500	250	25	500
3 Knapsacks	200	20	1000	250	25	1500	300	30	2000
4 Knapsacks	250	25	2500	300	30	3000	350	35	3500

6.3   Experimental Results

The mean coverage metrics of MOHEDA and MOGLS for nine test instances among 20 runs are given in Table 2. In the case of 2 knapsacks, there are nearly 98% of the solutions of MOGLS covered by the solutions of MOHEDA. It can be observed from Fig.1 that the nondominated solutions of MOHEDA are well-distributed along the true Pareto-optimal front. For the instances with 3 knapsacks, MOHEDA still performs better than MOGLS, and about 65% of non-dominated solutions of MOGLS are dominated by the solutions of MOHEDA. MOHEDA is slightly superior to MOGLS for the instances with 4 knapsacks. Nevertheless, the coverage metric $\mathcal{C}$(MOGLS,MOHEDA) decreases as the number of knapsacks increases. This might be caused by the immaturity of the population. If we increase the number of evolutionary generations, a better coverage metric could be obtained.

Table 2. Coverage Metrics for 9 Test Instances

Mean coverage	$\mathcal{C}$(MOHEDA,MOGLS)			$\mathcal{C}$(MOGLS,MOHEDA)		
Items	250	500	750	250	500	750
2 knapsacks	0.01%	0.00%	0.02%	99.11%	98.61%	97.92%
3 knapsacks	6.05%	2.62%	4.09%	67.20%	67.14%	61.53%
4 knapsacks	3.61%	5.16%	4.92%	30.90%	11.60%	9.13%

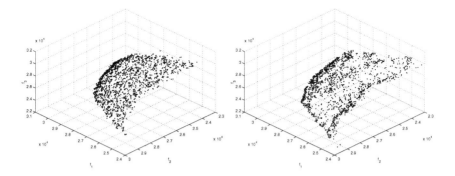

Fig. 2. Nondominated solutions of the instance with 3 knapsacks and 750 items: MO-HEDA(left), MOGLS(right).

Table 3. Convergence and Diversity Metrics

Performance Metrics		Convergence Metric ($\times100$)			Diversity Metric ($\times100$)		
Algorithms	Instances	Minimum	Mean	Maximum	Minimum	Mean	Maximum
MOGLS	(2,250)	0.1571	0.6115	1.1477	1.5270	2.7060	4.0700
	(2,500)	0.3031	1.4220	2.6603	2.6020	4.3010	5.4600
	(2,750)	1.0110	2.4612	4.3301	4.3920	5.5090	7.2640
MOHEDA	(2,250)	0	0.0514	0.2274	4.5620	7.0230	9.1780
	(2,500)	0.0216	0.1745	0.4071	6.4270	10.2030	12.5360
	(2,750)	0.1892	0.5420	1.6539	5.3970	11.0930	14.3690

We visualize the nondominated fronts in the case of two and three objectives. It can be intuitively observed from Fig.1 and Fig.2 that the nondominated fronts of MOHEDA are superior to those of MOGLS in both convergence and diversity. Moreover, the mean quantitative metrics of convergence and diversity for the instances with 2 knapsacks in 20 runs are given in Table 3. Here, we only give the results in the cases of 2 knapsacks in which the true or relaxed Pareto fronts are known to us. For all nine test instances, the values of the diversity metric of MOHEDA are greater than those of MOGLS. A better spread of nondominated front is achieved by MOHEDA. Obviously, from the convergence metrics in Table 3, the nondominated fronts of MOHEDA are closer to the Pareto-optimal front than those of MOGLS.

7 Conclusions

MOHEDA has been compared with MOGLS on MOKP. The experimental results show that MOHEDA outperforms MOGLS both in convergence and in diversity. The following techniques are effectively used in MOHEDA: (1) random repair method is used; (2) random neighbourhood structure is constructed in local search; (3) stochastic clustering method is employed; (4) a mixture-based UMDA is introduced. The distinct feature of MOHEDA is that multiple possible weighted sum functions can be optimized in a single generation. Moreover, the further work may focus on the performance analysis of MOHEDA.

References

1. A. Jaszkiewicz. On the Performance of Multiple-Objective Genetic Local Search on the 0/1 Knapsack Problem-A comparative Experiment, IEEE Transactions on Evolutionary Computation, Vol.6, No.4, August 2002.
2. Coello Coello, Carlos A. An Updated Survey of GA-Based Multiobjective Optimization Techniques. Technical Report Lania-RD-98-08, Laboratorio Nacional de Informática Avanzada (LANIA), Xalapa, Veracruz, M éxico, december 1998.
3. D. Thierens and P. A.N. Bosman. Multi-Objective Mixture-based Iterated Density Estimation Evolutionary Algorithms, Proceedings of the Genetic and Evolutionary Computation Conference (GECCO'2001) (2001).
4. E. Zitzler and L. Thiele. Multiobjective Evolutionary Algorithms: A Comparative Case Study and the Strength Pareto Approach. IEEE Transactions on Evolutionary Computation, 3(4), pages 257-271, November 1999.
5. H. Ishibuchi and T. Murata. Multi-objective Genetic Local Search Algorithm, in Proceedings of 1996 IEEE International Conference on Evolutionary computation(ICEC'96), iscataway, NJ, May 20-22 1996, IEEE, pp. 119-124.
6. H. Mühlenbein and G. Paaß. From Recombination of Genes to the Estimation of Distributions I. Binary parameters. In Lecture Notes in Computer Science 1411: Parallel Problem Solving from Nature-PPSN IV,178-187, 1996.
7. J. Knowles and D. Corne. M-PAES: A Memetic Algorithm for Multiobjective Optimization, in Proceeding of 2000 Congress on Evolutionary Computation. Piscataway, NJ:IEEE Press, July 2000, vol. 1,pp. 325-332.
8. M. Guntsch and M. Middendorf. Solving Multi-criteria Optimization Problems with Population-Based ACO, EMO 2003, LNCS 2632, pp. 464-478.
9. Laumanns Marco, Ocenásek Jirí. Bayesian Optimization Algorithms for Multi-Objective Optimization, In: Parallel Problem Solving from Nature - PPSN VII, Granada, ES, Springer, 2002, p.298-307, ISBN 3-540-444139-5.
10. P., Larrañaga, and J.A., Lozano. Estimation of Distribution Algorithms: A New Tools for Evolutionary Computation. Kluwer Academic Publishers, 2001.
11. Q. Zhang, J. Sun, E. P. K. Tsang, J. A. Ford. Hybrid Estimation of Distribution Algorithm for Global Optimisation, accepted in Engineering computations, 2003.

On the Use of Path Relinking for the
p – Hub Median Problem

Melquíades Pérez Pérez, Francisco Almeida Rodríguez, and
J. Marcos Moreno Vega

Dpto. de E.I.O. y Computación
Escuela Técnica Superior en Ingeniería Informática
Universidad de La Laguna
Facultad de Matemáticas y Física , 4ª planta.
Avda. Astrofísico Francisco Sánchez s/n
38271 La Laguna
Santa Cruz de Tenerife (SPAIN)
{melperez, falmeida, jmmoreno}@ull.es

Abstract. The *p – hub* median problem is an *NP* hard location – allocation problem, which consists of finding *p* points to establish facilities and the assignment of the users to these points. A new evolutionary approach that has been very effective for solving optimisation problems is Path Relinking, an extension of Scatter Search that links solutions over neighborhood spaces. Path Relinking gives a framework for combining solutions that goes beyond the crossover mechanism of genetic algorithms, and introduces new fundamental principles, such as the use of systematic strategies instead of random strategies. In this paper, we present an application of Path Relinking to solve the *p – hub* median problem and compare its effectiveness with other classical techniques. This procedure provides high quality solutions in reasonable execution times and yields a significant improvement in the size of the problems that can be solved.

1 Introduction

There are many real situations where several nodes must interact with each other by sending and receiving traffic flow of some nature. The flow exchanged may represent data, passengers, merchandise, express packages, etc. Generally, in situations like these, it is useful to find an optimal location of several points called hubs. The hubs act like exchanging points. The objective is to find the location of the hubs and to allocate the non hub nodes to the hubs minimizing an objective function that describes the interchanged flow and its cost. The hubs are fully interconnected and the traffic between any pair of nodes is routed through the hubs.

This problem was first formulated, by O'Kelly [10], as an integer quadratic problem. We will consider the *p-hub* median problem without capacities and with single assignment. We introduce the following notation.

Let *n* be a set of demand points. We denote, for any pair of nodes *i* and *j*,

J. Gottlieb and G.R. Raidl (Eds.): EvoCOP 2004, LNCS 3004, pp. 155–164, 2004.
© Springer-Verlag Berlin Heidelberg 2004

$$W_{ij} = \text{number of units sent from } i \text{ to } j$$
$$C_{ij} = \text{standard cost per unit of flow from } i \text{ to } j$$

Normally, $W_{ii} = 0$ and $C_{ii} = 0$, for all i, although there are some cases with $W_{ii} \neq 0$ and/or $C_{ii} \neq 0$. If one of the points, i or j, is a hub, the standard cost per unit C_{ij} is assumed. If they are both hubs, then the standard cost per unit of traffic is normally reduced and it is equal to αC_{ij}, where α is a parameter. In general, $\alpha \leq 1$ to reflect the effect of the reduced cost in inter–hubs flows. We can also consider parameters δ and γ to capture the influence of reduced or incremented costs for non hub points. There are not constraints for the parameters δ and γ.

Let X_{ij} and Y_j be the decision variables defined as:

$$X_{ij} = \begin{cases} 1 & \text{if point } i \text{ is allocated to hub } j \\ 0 & \text{otherwise} \end{cases} \qquad Y_j = \begin{cases} 1 & \text{if point } j \text{ is a hub} \\ 0 & \text{otherwise} \end{cases}$$

Then, the *p-hub* median location problem can be formulated as:

$$minf(x) = \sum_i \sum_j W_{ij} \times \left(\delta \sum_k X_{ik} C_{ik} + \gamma \sum_m X_{jm} C_{jm} + \alpha \sum_k \sum_m X_{ik} X_{jm} C_{km} \right)$$

subject to:

$$X_{ij} \leq Y_j \qquad \text{for i} = 1,\dots, \text{n and j} = 1,\dots, \text{n} \qquad (1)$$

$$\sum_j X_{ij} = 1 \qquad \text{for i} = 1,\dots, \text{n} \qquad (2)$$

$$\sum_j Y_j = p \qquad (3)$$

$$X_{ij}, Y_j \in \{0, 1\} \qquad \text{for i} = 1,\dots, \text{n and j} = 1,\dots, \text{n} \qquad (4)$$

Constraint (1) ensures that a non hub point is allocated to a location j if a hub is located in that site. Constraint (2) guarantees that each point is served by one and only one hub. And finally, constraint (3) generates the correct number of hubs. Constraint (4) is the classical binary constraint.

The *p-hub* median problem is an *NP*–hard problem. Furthermore, even if the hubs are located, the problem of allocating points is also *NP*–hard. Therefore, it is necessary to have recourse to heuristic solution approaches, especially when the problem contains a very large number of nodes.

In this paper, we present a new Path Relinking approach to solve the *p–hub* median problem. This new proposal uses systematic neighborhood-based strategies to explore the feasible region of the search space and provides good results even with large sizes problems. In section 2 we present the state of the art of the algorithms proposed for this problem. Section 3 introduces the principles of the Path Relinking method. Section 4 describes the implementation of the Path Relinking algorithm that we apply to solve the problem. Finally, Section 5 reports computational results disclosing that our algorithm constitutes a useful alternative approach for solving this problem.

2   The State of the Art

Several heuristics have been proposed for the *p-hub* problem. They can be classified attending to the phase they develop first. Some of them are focused on the study of the locational part and others on the assignment part. The first method to solve the *p-hub* problem was proposed by O'Kelly [10], based on an exhaustive search with two different assignment methods, HEUR1 and HEUR2. They enumerate all the possible location for hubs. The first one (HEUR1) assigns every demand point to the nearest hub on each hub configuration. The second one (HEUR2) assigns the non hubs to the nearest or to the second nearest hub. This work introduces a set of problem instances that have been used in the literature as test problems (CAB). Klincewicz [6] proposes more sophisticated heuristics. Foremost, he considers the traffic among the hubs in addition to the distance criteria. These two criteria are combined into a multicriteria assignment procedure. Klincewicz also considers exchange heuristics, taking into account only the configurations that yield immediately promising improvements, and proposes an associated *clustering* heuristic. Klincewicz [7] also investigates Tabu Search and GRASP. The heuristics proposed by Klincewicz provide running times better than the running times obtained with O'Kelly's proposals. A version of Tabu Search, called TABUHUB, is suggested in [12]. This heuristic uses a tabu search procedure for the location and allocation phases. The computational experience shows that TABUHUB is better than the O'Kelly's heuristics, HEUR1 and HEUR2. TABUHUB does not run as fast as Klincewicz's heuristics, but produces higher quality solutions and constitutes one of the best known heuristics to solve the *p–hub* median problem. M. Pérez et al. [9] present a hybrid method that combines a genetic algorithm with multistart search. The main contribution is a new encoding of the solutions and an associated selection of the operators, to produce a computational behaviour that is, sometimes, better than TABUHUB. Krishnamoorthy et al. [1] present new formulations reducing the number of variables and restrictions of the problem. They develop an algorithm based on Simulated Annealing and use the upper bound provided by this heuristic to create a Branch & Bound solution method. Computational results for the O'Kelly's problems and for a new set of problems (called the AP set) are presented. No comparisons with other heuristic approaches are available.

Some of the heuristics mentioned display a good behaviour in particular cases but, when the size of the problem increases, the heuristic performance decreases. Nowadays, the development of fast and efficient heuristics to solve large instances of the *p-hub* problem remains a challenging problem. We present here an adaptation of the Path Relinking to this problem. This new heuristic sets a remarkable improvement in the size of the problems solved while maintaining reasonable running times.

3   Path Relinking Principles

The Path Relinking (PR) method is an evolutionary metaheuristic procedure founded on the Scatter Search design originally formulated in Glover [3] in the 70s, but replacing the Euclidean spaces used in Scatter Search with neighbourhood spaces as a basis for combining solutions (Glover [4], Glover, Laguna and Marti [5]). Nowadays the literature contains many different applications of the procedure to hard optimization problems (see, e.g., the survey [5]). The method operates on a set of solutions called the *Reference Set, RS*, using strategies to select, combine, improve and generate new and better solutions from this set.

The Reference Set is a collection of the $|RS|$ more representative solutions of the problem, selected from an initial set of trial solutions, P. (Usually, the value of $|RS|$ is not larger than 20.) These solutions evolve to provide an intelligent exploration of the feasible region, through the schemes of intensification and diversification, which respectively focus the search on promising regions inferred from previous solutions and drive the search to explore new regions not previously visited.

Path Relinking applies the mechanisms of diversification and intensification to update the Reference Set and to guide the heuristic towards an optimal solution of the problem. The diversification of Path Relinking is customarily applied by updating a subset of the Reference Set with solutions not in the Reference Set that maximize the distance to the closest member of the Reference Set. The remaining solutions of the Reference Set are updated through the intensification scheme. Usually, the intensification is executed by combining different solutions to create new initial solutions for the local search procedures.

PR uses a mechanism for combining solutions that creates one or more neighbourhoods for moving from one solution to another, in the manner of neighbourhood search. Then PR executes a sequence of moves in the selected neighbourhood(s), to generate intermediate solutions that constitute the combinations produced (by the PR interpretation of the meaning of "combination"). In this way, the neighbourhood structure characteristically produces new solutions that contain some mix of the attributes of the parents, while preserving desirable properties such as feasibility by means of the neighbourhood definition. Multiple parents can be used simultaneously by constructing a neighbourhood path from one parent that at each step seeks to incorporate attractive attributes of other parents, where attractiveness is gauged by a selected evaluation criterion.

The success of Path Relinking is based on a good integration of the methods to select, combine and improve the solutions that update the Reference Set at each step of the procedure. The selection is carried out by partitioning the Reference Set into components. Here we use a 2-component partitioning [5] that divides the Reference Set in two subsets RS_1 and RS_2, respectively ($|RS_1| + |RS_2| = |RS|$). The first subset is composed of the $|RS_1|$ best solutions. The $|RS_2|$ most diversified solutions are included on the second subset. Then, two types of selection are performed. The solutions obtained by the combination and improvement are selected to update the first subset. The second subset is updated with the solutions in P more distant from the Reference Set.

The determination of the solutions to combine within the Reference Set is performed by generating subsets of the Reference Set. The solutions of these subsets are combined to obtain new solutions that incorporate good properties of the previous solutions. The subsets can be generated, for example, as follows:

- All the subsets of 2 solutions,
- All the subsets of 3 solutions derived from the 2-solutions subsets.
- All the subsets of i solutions derived form the (i-1)-solutions subset.

After that, the solutions obtained with the combination are improved with a simple local search.

In short, PR performs a systematic selection over a relatively small set of solutions of the Reference Set. Then, intensification is applied to improve the quality of the solutions through local search, and diversification is applied to enhance the dispersion of the solutions considered. New solutions are generated by drawing on the information about the quality of known solutions.

The scheme in Figure 1 shows a basic outline of the procedure. This scheme uses the following variables:

- P: initial set of diverse solutions
- Ref_Set: set of reference solutions,
- x: current solution and
- x^*: improved solution.

```
procedure Path Relinking
    begin
            PopulationCreation(P) ;
            ReferenceSetGeneration(RefSet) ;
            While (ReferenceSet is updated OR number of iterations) do
                        Repeat
                                    SubsetGeneration(Subset) ;
                                    SolutionCombination(Subset , x) ;
                                    Improvement (x , x*) ;
                                    ReferenceSetUpdate ;
                                    end ;
                        until (stopping rule) ;
            End While ;
    End
```

Fig. 1. Outline of the *Path Relinking*

4 Path Relinking for the *p – Hub* Median Problem

4.1 Encoding Solution

We present our representation for the solutions of the *p–hub* problem and its utilization within the *PR* approach. The matrix with restrictions (1), (2), (3) and (4) is a natural representation of a solution. Nevertheless, this representation is difficult to use when designing efficient combination methods. Then, we propose the following representation:

Each solution is represented by an array of size n that indicates the hubs and the assignment of the rest of nodes to these hubs. Lets assume that the set of demand points is indexed by the set $L=\{1, \dots , n\}$, the idea of our representation is the following: the first p positions of the array, called *location–array*, are the ordered indexes of the hubs; and, the last $(n – p)$ positions, called *allocation–array,* designate the allocations of the non hub nodes, assuming that the hubs are allocated to themselves.

Suppose, for example, that $n = 5$, $p = 2$, and $L = (1\ 2\ 3\ 4\ 5)$. In this context, the solution X, where hubs are the nodes 3 and 4, and the first node is assigned to hub 4, and the second and the fifth nodes are assigned to hub 3, is represented with the matrix:

$$X = \begin{pmatrix} 0 & 0 & 0 & 1 & 0 \\ 0 & 0 & 1 & 0 & 0 \\ 0 & 0 & 1 & 0 & 0 \\ 0 & 0 & 0 & 1 & 0 \\ 0 & 0 & 1 & 0 & 0 \end{pmatrix}$$
is represented by the following array: $s = \begin{pmatrix} 3 & 4 & | & 2 & 1 & 1 \end{pmatrix}$

4.2 Initialization Method

The initial population of solutions is generated using a Multistart method. The method operates as follows:

1 Generate a random solution and improve it with a greedy procedure in the location phase and a greedy procedure in the allocation phase. The greedy procedure in location was used previously in [8] with the name of LS2. On this greedy procedure, every solution of the *p-hub* problem induces a partition over the set of demand points A set of the partition is constituted by a hub and the non hub points assigned to it. The neighbourhoods used to apply the greedy mechanism, are obtained through the substitution of a hub i by every point of the set of the partition associated to i, and assigning the non hub points to the nearest hub. This procedure is repeated until not improvements are achieved. Then, a greedy procedure is applied for the assignment. The greedy assignment procedure consists of exchange every non hub j from its current assignment to a different one until no improvements. If the greedy assignment procedure improves the solution, we proceed again with the new neighbourhood. At the end of the greedy procedure, only new solutions are introduced in P. A pseudocode for this procedure is illustrated in Figure 2.

2 If the improved solution does not belong to P, P is updated with this new solution.

3 The steps 1 and 2 are repeated until the elements in P are P_size.

4.3 Reference Set Generation Method

RS is initialized with the best b_1 solutions of P. Then, the b_2 solutions in $P - RS$ that are the most disparate (dispersed) with respect to RS are added to RS. The solution t is defined to be the most disparate solution with respect to the solutions already in RS, if t maximizes the distance function:

$$d(t, RS) = \min\{d(t, s) / s \in RS\}$$

with: $d(t,s) = \rho \cdot d\_loc(t, s) + (1 - \rho) \cdot d\_as(t, s)$

where: $d\_loc(t, s)$ = number of different hubs in s and t

$d\_as(t, s)$ = number of nodes with different assignment in both solutions

4.4 Subset Generation Method

We consider all the subsets of 2 solutions of the reference set. On this way, the stopping rule of the pseudocode (Figure 1) is the total enumeration of all the subsets with two solutions.

4.5 Relinking Method

The relinking between each pair of solutions into each subset of the reference set is achieved with the neighbourhood structure defined below. Moves within these neighbourhoods determine the location relinking and leave the allocation relinking to a random strategy. The underlying idea is that if we try to preserve the common information in location, the allocation relinking phase will be dominated by the location relinking.

Location Relinking Neighbourhood: This neighbourhood provides moves to build a path between two selected solutions. The application of the *Improvement Method* defined below, will create new solutions from the solutions of this path. Each solution of the path will be generated swapping a random element on both the initial and the final solution of the path. The moves of the neighbourhood are structured so that they yield feasible outcomes at each step. In particular, if we consider an instance of the problem with $n = 15$ and $p = 6$, and if we take the solutions with location arrays $l_i = (2\ 7\ 9\ 10\ 11\ 15)$, and $l_f = (1\ 3\ 8\ 5\ 9\ 10)$, then the path is produced as follows:

- marks the common hubs of l_1 and l_2: $l_1 = (2\ 7\ \underline{9}\ \underline{10}\ 11\ 15)$ $l_2 = (1\ 3\ 8\ 5\ \underline{9}\ \underline{10})$
- places on the first positions (ordered) the non marked symbols:

$$l_1 = (2\ 7\ 11\ 15\ \underline{9}\ \underline{10})$$
$$l_2 = (1\ 3\ 5\ 8\ \underline{9}\ \underline{10})$$

- the intermediate solutions on the path will be:

$$l_1 = (1\ 7\ 11\ 15\ 9\ 10) \qquad l_2 = (1\ 3\ 11\ 15\ 9\ 10) \qquad l_3 = (1\ 3\ 5\ 15\ 9\ 10)$$

The feasibility of the solutions at the end of the location relinking method, is ensured with the use of the encoding we have defined above.

Allocation Relinking Neighbourhood: This neighbourhood is defined by reference to moves of the following type. For each pair of allocation arrays, a_1 and a_2, generate a random position of the range $[1\ \dots\ n - p - 1]$, and simultaneously execute a move from a_1 towards a_2 and from a_2 towards a_1 by swapping the two tail portions. For example, if we consider the instance used in the location relinking and if the two allocation–arrays to be operate are $a_1 = (1\ 1\ 2\ 3\ 1\ 5\ 2\ 3\ 5\ 4)$ and $a_2 = (2\ 1\ 3\ 1\ 4\ 2\ 5\ 5\ 2\ 3)$, the new solutions are generated with the following process:

- generate a random position: $a_1 = (1\ 1\ 2\ 3\ 1\ 5\ |\ 2\ 3\ 5\ 4)$, $a_2 = (2\ 1\ 3\ 1\ 4\ 2\ |\ 5\ 5\ 2\ 3)$
- switch the two tail portions: $a_1 = (1\ 1\ 2\ 3\ 1\ 5\ |\ 5\ 5\ 2\ 3)$, $a_2 = (2\ 1\ 3\ 1\ 4\ 2\ |\ 2\ 3\ 5\ 4)$

4.6 Improvement Method

As an improvement method, we use a greedy procedure both for location and allocation. This kind of procedure is commonly used in the Scatter Search technique. The greedy procedure used in this case has been defined in 4.2. and was formulated previously in [8], where it was proved to be very effective. The improvement method is applied only to a subset of the solution on the path. This subset is formed by the solutions with the large incoming and outcoming flow.

4.7 Reference Set Update Method

At the end of every iteration, each solution within the path created as above is compared with the worst solution within the *RS* (attending to the objective value) and if improves it, the *RS* is updated with that solution.

5 Computational Experience

The computational experience compares the behaviour of the Path Relinking method for the *p-hub* problem, against the Tabu Search procedure TABUHUB. The algorithms have been implemented using C language programming and they have been executed on a 1400 Mhz Intel Xeon PC, with 1 Gb of RAM, running the Linux Operating System.

Our testing is performed over the AP Problem set (see [2]), corresponding to the Australian postal office between 200 major cities. For these problems, the optimal objective values for n ≤ 50, and the best known solution for n > 50, can be found in [1]. Our tables present the computational results obtained for problems with $n = 10$, 20, 25, 40 and 50, and $p = 2, 3, 4$ and 5. For $n = 100$, $p = 5, 10, 15, 20$.

Since TABUHUB and *PR* are random heuristics, they were executed 30 times on each problem. For each problem, the tables show the optimal objective value, and the results for both heuristics (time measured in minutes and seconds, and the factor ϕ – in percent – by which the objective function exceeds the best known solution):

$$\phi = 100 \cdot \left(\frac{\text{objective - best known solution}}{\text{best known solution}} \right)$$

In both tables below, the column labelled TABUHUB presents the results obtained for the TABUHUB procedure and the columns labelled *PR* give the results of the Path Relinking method. For both heuristics minimum *(Min)*, average *(μ)* and maximum *(Max)* values on the 30 executions are showed. The column labelled %, represents the number of times (on percentage), that the heuristic reached the best known solution. On the *PR*, the size of the population was $P = 10$, and the $|RS| = b = 10$. These sizes were selected experimentally. The TABUHUB algorithm has been executed using the running parameters given in [12]. The number of iterations of the *PR* procedure were fixed between 5 and 10. The AP problems were chosen to yield a test set containing large problem instances.

As the tables below show, the Path Relinking procedure provides, in general terms, a better behaviour than TABUHUB. The results obtained by both heuristics for small problems are very similar, but when the number of nodes increases, the new *PR* approach improves the results of TABUHUB both in quality of solutions and execution times. In addition, the execution times of TABUHUB have an exponential behaviour while the times of our Path Relinking procedure seem to have a linear performance.

Table 1. Computational Results for the AP problem

			TABUHUB				PR			
				φ				φ		
					CPU time				CPU time	
n	p	Optimal	%	Min	μ	Max	%	Min	μ	Max
10	2	167493.06	100	0.00	0.00	0.00	100	0.00	0.00	0.00
				0.15	0.25	0.40		0.00	0.01	0.02
	3	136008.13	100	0.00	0.00	0.00	100	0.00	0.00	0.00
				0.50	0.65	1.20		0.00	0.01	0.02
	4	112396.07	100	0.00	0.00	0.00	100	0.00	0.00	0.00
				1.45	1.55	2.00		0.00	0.01	0.02
	5	91105.37	100	0.00	0.00	0.00	100	0.00	0.00	0.00
				2.40	2.43	2.55		0.00	0.01	0.02
20	2	172816.69	100	0.00	0.00	0.00	100	0.00	0.00	0.00
				1.59	1.60	1.70		0.00	0.04	0.06
	3	151533.08	97	0.00	0.00	0.15	100	0.00	0.00	0.00
				2.09	2.12	2.15		0.00	0.04	0.08
	4	135624.88	100	0.00	0.00	0.00	100	0.00	0.00	0.00
				2.06	2.20	2.45		0.00	0.06	0.10
	5	123130.09	100	0.00	0.00	0.00	100	0.00	0.00	0.00
				5.28	5.35	6.12		0.00	0.10	0.12
25	2	175541.98	100	0.00	0.00	0.00	100	0.00	0.00	0.00
				2.42	2.59	3.12		0.00	0.09	0.12
	3	155256.32	100	0.00	0.00	0.00	100	0.00	0.00	0.00
				5.43	6.05	6.15		0.00	0.12	0.22
	4	139197.17	90	0.00	0.00	0.20	100	0.00	0.00	0.00
				8.42	8.59	9.10		0.00	0.15	0.30
	5	123574.29	97	0.00	0.00	0.15	97	0.00	0.00	0.06
				15.35	15.48	16.05		0.00	0.18	0.30
40	2	177471.67	100	0.00	0.00	0.00	100	0.00	0.00	0.00
				4.23	4.36	4.40		0.00	0.28	0.40
	3	158830.54	100	0.00	0.00	0.00	100	0.00	0.00	0.00
				6.42	6.50	6.59		0.00	0.59	1.15
	4	143968.88	90	0.00	0.00	0.15	100	0.00	0.00	0.00
				12.28	12.40	12.55		0.00	1.18	1.22
	5	134264.97	90	0.00	0.00	0.10	90	0.00	0.00	0.10
				18.40	19.01	19.20		0.00	1.26	1.38
50	2	178484.29	80	0.00	0.01	0.10	100	0.00	0.00	0.00
				5.48	5.58	6.10		0.10	1.12	1.22
	3	158569.93	90	0.00	0.00	0.15	97	0.00	0.00	0.18
				6.32	6.43	6.50		0.10	2.16	2.30
	4	143378.05	80	0.00	0.01	0.20	90	0.00	0.00	0.09
				16.10	16.45	17.02		0.40	3.19	4.30
	5	132366.95	70	0.00	0.11	0.35	97	0.00	0.00	0.15
				20.25	20.40	20.55		0.40	3.24	4.42
100	5	136929.44	70	0.00	0.10	0.15	90	0.00	0.00	0.09
				12.50	13.25	15.50		0.40	5.01	7.20
	10	106469.57	50	0.00	0.18	0.22	80	0.00	0.05	0.10
				20.22	20.59	24.20		0.40	6.36	8.55
	15	90605.10	50	0.00	0.12	0.15	80	0.00	0.10	0.15
				23.59	24.40	28.54		1.20	9.50	13.58
	20	80682.71	30	0.00	0.26	0.34	70	0.00	0.12	0.18
				30.25	38.50	40.26		2.50	16.48	20.50

In conclusion, our new Path Relinking metaheuristic for the *p-hub* problem performs better than TABUHUB overall, and performs significantly better than TABUHUB on large problem instances. We anticipate the possibility of further improvement of our procedure by additional refinement of the neighbourhoods used to define moves for combining solutions. In particular, our current neighbourhoods are designed for executing a single large step move in each, to transform the parents into offspring for the next iteration. Neighbourhoods that transition between the solutions using additional intermediate steps are also appropriate to consider.

Acknowledgments. The third author has been partially supported by the Spanish Ministry of Science and Technology through the project TIC2002-04242-C03-01; 70% of which are FEDER founds.

References

1. Andreas T. Ernst and Mohan Krishnamoorthy, Efficient Algorithms for the Uncapacitated Single Allocation *p-hub* Median Problem. Location Science, vol. 4, No.3, (1995), 130 – 154.
2. Beasley, J.E. "OR-library" http://mscmga.ms.ic.ac.uk/info.html.
3. Glover, F. Heuristics for Integer Programming Using Surrogate Constraints. Decision Sciences, Vol 8, No 1, (1977), 156 – 166.
4. Glover, F. Tabu Search for Nonlinear and Parametric Optimization (with Links to Genetic Algorithms), Discrete Applied Mathematics, 49 (1994), 231-255.
5. Glover, F., M. Laguna and R. Marti. Fundamentals of Scatter Search and Path Relinking. Control and Cybernetics, Vol. 29, No. 3 (2000), 653-684.
6. Klincewicz, J.G. Heuristics for the *p-hub* location problem. European Journal of Operational Research, Vol. 53 (1991), 25 – 37.
7. Klincewicz, J.G. Avoiding local optima in the *p-hub* location problem using tabu search and grasp. Annals of Operation Research, Vol. 40 (1992), 283 – 302.
8. M. Pérez, F. Almeida, J. M. Moreno – Vega. Fast Heuristics for the *p-hub* Location Problem, presented in EWGLAX, Murcia (Spain), (1998).
9. M. Pérez, F. Almeida, J. M. Moreno – Vega. Genetic Algorithm with Multistart Search for the *p-Hub* Median Problem. Proceedings of EUROMICRO 98. IEEE Computer Society (2000), 702-707.
10. O'Kelly, M. (1986). The location of interacting hub facilities, Transportation Science, Vol 20, (1986), 92 – 106.
11. Skorin-Kapov, D & Skorin-Kapov, J. (1994). "On tabu search for the location of interacting hub facilities". European Journal of Operations Research, Vol. 73, 502-509.
12. Ernst, A.T., H. Hamacher, H. Jiang, M. Krishnamoorthy and G. Woeginger (2002). Uncapacitated Single and Multiple Allocation *p-Hub* Center Problems", to appear.

Solving a Real-World Glass Cutting Problem

Jakob Puchinger, Günther R. Raidl, and Gabriele Koller

Institute of Computer Graphics and Algorithms,
Vienna University of Technology, Vienna, Austria
{puchinger,raidl,koller}@ads.tuwien.ac.at

Abstract. We consider the problem of finding two-dimensional cutting patterns for glass sheets in order to produce rectangular elements requested by customers. The number of needed sheets, and therefore the waste, is to be minimized. The cutting facility requires three-staged guillotineable cutting patterns, and in addition, a plan for loading produced elements on transportation wagons. The availability of only three loading docks for wagons imposes additional constraints on the feasibility of cutting patterns. We describe a greedy first fit heuristic, two branch-and-bound based heuristics, and different variants of an evolutionary algorithm for the problem, and compare them on a set of real-world instances. The evolutionary algorithm is based on an order representation, specific recombination and mutation operators, and a decoding heuristic. In one variant, branch-and-bound is occasionally applied to locally optimize parts of a solution during decoding.

1 Introduction

In this paper we present methods developed for solving a cutting problem that appears in real-world glass manufacturing. Customer specified rectangular elements need to be cut out from raw stock sheets. Due to the mechanics of the cutting machine, only three-staged guillotine cuts are possible, and additional constraints regarding the order in which elements are cut, must be obeyed. The problem falls into the general category of two-dimensional bin packing (2BP) problems; according to the notation of Lodi et al. [10] it can be classified as 2BP|R|G with additional constraints.

Our problem is enhanced by constraints resulting from the characteristics of the production process. A robot located at the end of the glass-cutting facility puts finished elements on wagons located at three different loading docks. The elements are partitioned in logical groups according to customer orders, and each wagon may carry elements from the same logical group only. Since the wagons cannot be exchanged arbitrarily, elements from a maximum of three logical groups can be produced in an intertwined way. A solution to our problem must also contain a schedule for loading finished elements onto wagons.

A detailed problem description is given in Sec. 2. Related methods from literature for solving the classical two-dimensional bin packing problem are summarized in Sec. 3. Sections 4 and 5 present a new greedy heuristic and two branch

This work is supported by the Austrian Science Fund (FWF) under grant P16263-N04.

J. Gottlieb and G.R. Raidl (Eds.): EvoCOP 2004, LNCS 3004, pp. 165–176, 2004.
© Springer-Verlag Berlin Heidelberg 2004

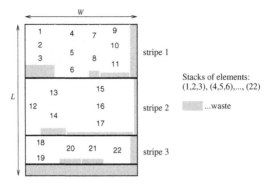

Fig. 1. An example for a three-stage cutting pattern in normal form.

and bound algorithms for our problem. Section 6 describes three variants of an evolutionary algorithm (EA). The different algorithms are experimentally evaluated and compared to each other and to a commercial solver in Sec. 7. In most cases, the EA yields the best results. More details on this work are given in [14].

2   Problem Description

In the two-dimensional three-staged Bin Packing problem with Wagon Scheduling (2BP-WS), we are given:

- Identical rectangular stock sheets of width $W > 0$, length $L > 0$, and thickness $T > 0$.
- A set of rectangular element types $\mathcal{E} = \{E_1, \ldots, E_m\}$ requested by customers; each element type E_i, $i = 1, \ldots, m$, is defined by the elements' width w_i ($0 < w_i \leq W$), length l_i ($0 < l_i \leq L$), a logical group identifier g_i indicating the customer to whom the elements are to be sent, and the number $n_i > 0$ of instances needed of this type. Let $n = \sum_{i=1}^{m} n_i$ be the total number of elements.
- A minimum fill level $B_{\min} > 0$ and a maximum fill level $B_{\max} \geq B_{\min}$ for the transportation wagons; the fill level of a wagon is given by the cumulated thickness of the elements loaded on it.

The goal is to find patterns for cutting out all instances of the given element types from the smallest possible number of stock sheets, thus, minimizing the waste, and a schedule for loading produced elements to wagons. Elements may be rotated by 90°. The production machine and loading robot impose the following restrictions on feasible cutting patterns.

Orthogonal guillotine cuts. Since glass sheets are cut by scratching and breaking them over a straight edge, only orthogonal guillotine cuts are possible, i.e. cuts parallel to a side of the piece being cut, going from one border straight to the opposite side.

Three stages. The production machine consists of three stages. The first stage only cuts a whole sheet horizontally into an arbitrary number of stripes. Stripes are further processed in the second stage where they are cut vertically into an arbitrary number of so-called stacks. The third stage performs, again, only horizontal cuts, producing the final elements (and waste) from the stacks. See Fig. 1 for an example of a three-stage cutting pattern.

Any feasible three-stage cutting pattern can always be reduced into its so-called norm al form by moving each element to its uppermost and leftmost position, so that waste may only appear at the bottom of the stacks on the right of the rightmost stack in each stripe or at the bottom below the last stripe of each sheet. The pattern shown in Fig. 1 is in normal form. In the rest of the paper, we only consider cutting patterns in normal form, which reduces the general infinite search space of arbitrary three-stage cutting patterns to a finite number of possible solutions.

Wagon restrictions. For a given cutting pattern, elements are always produced by processing the stripes within each sheet from top to bottom, the stacks within each stripe from left to right, and the elements within a stack from top to bottom. The numbering of the elements in Fig. 1 denotes this order.

Finished elements are loaded on wagons in the same order as they are produced. Initially, three empty wagons are placed at three available loading docks. The following constraints apply to the loading of elements on wagons.

1. The first element put on a wagon initializes the wagon's logical group, i.e. it defines its destination address. Only elements of the same logical group may further be added.
2. A wagon is immediately closed and replaced by a new, empty one, if the fill level of a wagon reaches B_{max} or if all elements of its associated logical group have been produced and loaded.
3. A wagon may be closed and replaced earlier if its fill level is at least B_{min}.

Only cutting patterns, for which a loading schedule fulfilling the above requirements exists, are considered feasible.

2.1   Objective Function and Continuous Lower Bound

The primary objective is to minimize the number $S(x)$ of sheets needed in a solution x. A refined objective function $F(x)$ considers the last sheet only partly:

$$F(x) = S(x) - \frac{L - c_l}{L}, \tag{1}$$

where c_l represents the position of the last stage-1 cut of the last sheet. This refined objective function allows for a better discrimination of the quality of solutions needing the same number of sheets: Cutting patterns leaving a larger unused stripe at the bottom of the last sheet are more promising. In practice, large unused final stripes may be useful later on and are therefore not necessarily considered as waste.

The continuous lower bound CLB for the objective function ($S(x)$ as well as $F(x)$) is given by the total area of all elements divided by a sheet's area:

$$\mathrm{CLB} = \frac{\sum_{i=1}^{m} l_i \cdot w_i \cdot n_i}{L \cdot W} \tag{2}$$

3   Previous Approaches to Two-Dimensional Bin Packing

In this section we give a brief overview of known algorithms for solving the two-dimensional bin packing problem on which the new algorithms are based. For a more complete review we refer to the survey by Lodi et al. [9] and to the annotated bibliography by Dyckhoff et al. [5].

Most of the simple algorithms for 2BP are of greedy nature. The elements are placed one by one and never reconsidered again. There are one- and two-phase algorithms. In one-phase algorithms, the elements are directly placed in the bins. Two-phase algorithms first partition the elements into levels whose widths do not exceed W and whose total length is aimed to be minimized. In the second phase, levels are assigned to bins by heuristically solving a one-dimensional bin packing problem.

Berkey and Wang [1] described the classical Finite First Fit (FFF) heuristic, which is a greedy one-phase algorithm. The elements are sorted by decreasing lengths. The first element initializes the first level in the first bin and defines the level's length. Each following element is added to the first level to which it fits respecting the sheet's width W. If there is no such level, a new level is initialized in the first bin into which it fits. If there is no such bin, a new bin is initialized with the new level. Within a level, elements are never stacked.

For the general 2BP problem, Martello and Vigo [11] describe a branch-and-bound (B&B) algorithm which is based on a two-level branching-scheme: The elements are assigned to the bins by an outer decision tree. Possible packing patterns for the bins are heuristically generated. Only if the heuristic is not able to place all the elements of a bin, a pattern is tried to be found by an inner enumeration scheme. Morabito and Arenales [13] consider a staged and constrained guillotine cutting problem in which the utilization of only a single sheet is maximized. They describe a B&B algorithm whose decision tree corresponds to an AND/OR graph.

A broad overview of EAs for cutting and packing problems is given by Hopper [6]. Most of the EAs used to solve guillotineable 2BP, such as the genetic algorithm described by Kröger [8], are based on a slicing-tree representation and specific variation operators. Alternatively, an order-based encoding can be used indicating the order in which the elements are placed by some FFF-like decoding heuristic. Hwang et al. [7] describe such an approach and compare it to a slicing-tree representation. They conclude that the order-based method yields better results in several cases. Corno et al. [3] apply the order-based encoding successfully to an industrial cutting problem with specific real-time and pattern constraints. Another effective order-based approach is presented by Monaci [12] for two-dimensional strip packing.

Lodi et al. [10] describe a more general framework applicable to several variants of 2BP. Tabu search is used to assign the elements to the sheets, and cutting patterns for individual sheets are obtained by different inner heuristics.

Conclusions with respect to 2BP-WS. The algorithms presented so far cannot directly be applied to 2BP-WS because of the constraints imposed by the required wagon scheduling. Furthermore, several algorithms do not consider the restriction to three stages. The B&B algorithms, in particular, rely heavily on several symmetries of the general 2BP problem: Bins within a solution, levels within a bin, and elements within a level can be arbitrarily permuted. This is exploited by only considering levels and elements in specific orders. In case of 2BP-WS this simplification is not possible because the production order is a substantial part of the solution. Order-based EAs, however, do not principally rely on certain symmetries and can be adapted to our needs by modifying the decoding heuristic.

4   A Finite First Fit Heuristic for 2BP-WS (FFFWS)

The FFF heuristic from [1] has been modified and extended for 2BP-WS. In contrast to the classical FFF heuristic and due to the wagon scheduling requirements, FFFWS constructs a solution sheet by sheet and stripe by stripe in a greedy way; i.e. once a new sheet (stripe) is started, previous sheets (stripes) are never reconsidered.

Let $\Pi = (\Pi_1, \ldots, \Pi_m)$ be a list of references to the element types sorted according to decreasing lengths l_i, and let $|\Pi_1|, \ldots, |\Pi_m|$ be the corresponding numbers of not yet placed elements; initially $|\Pi_i| = n_i$.

The algorithm starts by placing a longest available element, i.e. an element of type Π_1, at the upper left position of the first sheet. This element is said to initialize the sheet and the first stripe, and it also defines this stripe's length $\hat{l}$. Whenever an element is placed, the corresponding counter $|\Pi_i|$ is decreased; if $|\Pi_i| = 0$, Π_i is removed from the list Π. The stripe is then iteratively filled from left to right with a stack of elements from the first type Π_i in list Π fulfilling the following conditions: (a) $l_{\Pi_i} \leq \hat{l}$; (b) w_i does not exceed the stripe's remaining width; (c) a permitted loading dock exists for elements of type Π_i with respect to wagon scheduling (see below). The number of stacked elements is $\min(|\Pi_i|, \lfloor \hat{l}/l_{\Pi_i} \rfloor)$. If no further element fits onto the current stripe, a new stripe is started with an element of the first type Π_i in list Π for which l_{Π_i} does not exceed the sheet's remaining length and a permitted loading dock is available. If no such element exists and Π is not empty, a stripe on a new sheet is started in the same way.

FFFWS tries to determine a permitted loading dock for an element by considering the element's logical group g_i and the current states of the wagons at the three loading docks. If g_i appears as logical group of one of the wagons at the loading docks, then this loading dock is a permitted target. Otherwise, we look for a loading dock whose wagon is either empty or has reached its minimum fill level $B_{\min}$. In the latter case, the wagon is closed and replaced by a new one. If a permitted loading dock could be determined, its identifier is stored as loading target for the element as part of the solution. Wagon changes need not explicitly be stored in this kind of greedy solutions, since they can be implicitly derived.

The possible rotation of elements is considered by maintaining for each element type two references Π_i and Π_i' in Π with linked counters of not yet placed elements; hereby, Π_i' represents the rotated version of the element type.

In the worst-case, FFFWS has to process the complete list Π when looking for the next element to be placed. Therefore, the worst-case total time complexity of FFFWS is $O(n \cdot m)$.

5   Branch-and-Bound Algorithms for 2BP-WS

Two different B&B strategies were devised. Since symmetries cannot be exploited as in case of the classical 2BP, it seems to be impossible to solve 2BP-WS instances of practical size to provable optimality. Therefore, the following B&B strategies also contain heuristics and cannot guarantee optimal solutions.

BB-1. The first B&B variant follows the principle of the greedy two-phase algorithms for 2BP. In the first phase, BB-1 iteratively generates stripes of width W and length not exceeding L containing elements not yet placed on previous stripes. B&B is used to minimize the difference between the stripe's area and the total area of the elements on the stripe, i.e. the waste. Subproblems are created from partially filled stripes by considering any feasible extension by a single element or a stack of elements of the same type. A breadth-first enumeration strategy is applied. The waste of the currently best stripe is used as global upper bound. As local lower bound, the subproblem's waste reduced by the empty area to the right of the last stack is used. If during the construction of the stripe a wagon change involving also a change of the logical group occurs, we consider the stripe a delim iter.

In the second phase, stripes are arranged on sheets by an adapted first fit decreasing algorithm for one-dimensional bin packing [2]. Due to wagon scheduling, only the order of stripes in sections between delimiters may be permuted.

BB-2. In contrast to BB-1, BB-2 maximizes the utilized area of whole sheets instead of minimizing the waste of single stripes. The patterns for the sheets are generated one by one by B&B. New subproblems are created from partially filled sheets by considering any feasible extension by a single element or a stack of elements of the same type, if necessary on a new stripe. The FFFWS heuristic is applied to get an initial global lower bound for the utilized area. The local upper bound of a subproblem is calculated as the total area of already placed elements plus an upper bound for the area that might further be utilized. Additional elements may only be placed to the right of the last stack on the last stripe or below the last stripe. Again, a breadth-first enumeration strategy is applied.

6   Evolutionary Algorithms for 2BP-WS

The presented algorithms FFFWS, BB-1, and BB-2 yield respectable results in several cases. Nevertheless, independently optimized stripes or sheets, as in case of BB-1 and BB-2, do not guarantee an optimal overall solution. In order to approach the problem in a more global way, we apply evolutionary algorithms.

Representation and Decoding. Two variants of an order-based representation similar to those in [7,3] were chosen to encode candidate solutions.

Element type representation. Solutions are encoded as ordered vectors of references to element types $\Pi = (\Pi_1, \ldots, \Pi_m)$. The phenotypic solution is derived by applying a modified FFFWS heuristic as decoder. Instead of considering element types ordered by decreasing length, the order is given by the genotype. Furthermore, the decoding heuristic allows stacked elements to extend a stripe's length if the total waste of the stripe decreases.

Element representation. The element type representation has the disadvantage of placing elements of the same type always next to each other, if space permits. To allow more flexibility, we also consider an order-based representation in which a solution is encoded as permutation $\pi = (\pi_1, \ldots, \pi_n)$ of references to each individual element. This can significantly increase the size of the genotype and the solution space in general. The element type representation's decoding heuristic has been adapted accordingly. In particular, elements of the same type may only be stacked by the decoding heuristic if they appear directly next to each other in the encoded solution.

Rotation of elements is in both representations handled by maintaining linked second references as in FFFWS. For simplicity, we neglect here the resulting changes in the genotype's length.

Recombination and Mutation. In principle, any standard recombination and mutation operator designed for permutations is feasible. Regarding recombination, *order 3 crossover* (OX3) [4] turned out to be particularly well suited because it partially respects absolute positions and relative orders, which is a crucial condition for our problem.

Two mutation operators are used. *Reciprocal exchange* (RX) chooses two positions at random and swaps them. *Block exchange* (BX) swaps two randomly chosen non-overlapping blocks of length $\lceil 2^R \rceil$, with R being a random value in the interval $(0, \lfloor \mathrm{ld}\, \frac{m}{2} \rfloor]$ for the element type representation and $(0, \lfloor \mathrm{ld}\, \frac{n}{2} \rfloor]$ for the element representation. In this way, shorter blocks are chosen more likely, but large blocks up to half of the genotype's length are also possible.

Two additional operators were specifically designed for the element representation with the aim to encourage sequences of elements of the same type in order to exploit stacking more often.

Grouped order crossover (GOX) is an extension of the standard OX3 operator. As in OX3, two crossover points are randomly chosen. However, only positions inbetween elements of different types are allowed. The crossover region defined by the two crossover points is transmitted directly from the first parent

Algorithm Grouping Mutation (π);
$j \leftarrow$ a random value $\in \{1, \ldots, n\}$; $//$ starting position
$p \leftarrow$ a random value $\in [0, 1)$; $//$ element moving probability
$left \leftarrow j - 1$; $right \leftarrow j + 1$;
for $i \leftarrow j - 1, \ldots, 1$ **do**
   if type of element π_i = type of element π_j **then**
      with probability p **do**
         swap (π_i, π_{left}); $left \leftarrow left - 1$;
for $i \leftarrow j + 1, \ldots, n$ **do**
   if type of element π_i = type of element π_j **then**
      with probability p **do**
         swap (π_i, π_{right}); $right \leftarrow right + 1$;

Fig. 2. Grouping mutation operator for the element representation.

to the offspring. All remaining positions are filled with the remaining elements of the first parent in the order given by the second parent.

Grouping mutation (GM) is used to form larger sequences of elements of the same type. Figure 2 shows the algorithm in detail. A position j of the permutation π is randomly chosen as starting position. π is scanned and each element of the same type as π_j is incorporated into the sequence around j with a probability chosen randomly from $[0, 1)$.

The computational effort of both operators, GOX and GM, is linear in the number of elements.

EA Variants. The described operators were integrated in a standard steady-state EA. Initial solutions are created at random. In each iteration, a new candidate solution is created by always applying crossover and mutation. The new solution replaces the worst solution in the population if it is not identical to an already existing solution on the genotype level.

We consider the following three EA variants:

- EAet uses the element type representation.
- EAe uses the element representation.
- EAebb corresponds to EAe enhanced by phase 1 of BB-1: In the decoding, stripes are not just created by the greedy heuristic, but with a certain probability p_{bb} by applying B&B.

In case of EAe and EAebb the order of the elements in a decoded solution is reencoded into the genotype in a Lamarckian manner with probability p_{wb}.

7   Computational Results

31 real-world instances were provided by Soglatec GmbH and are available at http://www.ads.tuwien.ac.at/pub/TestProblems/glasscut. These instances have strongly varying characteristics: 1 to 31 logical groups, 1 to 78 element types, 4 to 651 elements in total, and continuous lower bounds CLB ranging from 0.34 to 23.90.

Table 1. Summarized results of selected algorithms; $\overline{CLB} = 6.010$; * denotes that the mutation operator was randomly chosen among RX, BX, and GM; the value printed after the mutation type represents p_{wb}; for EAebb the last value represents p_{bb}.

algorithm	$\overline{F}$	$\overline{\sigma_F}$	F/CLB	$\bar{t}[s]$	best
XOPTS	7.195	n.a.	1.1967	n.a.	15
FFFWS	7.896	n.a.	1.3191	0.02	2
BB-1	8.280	n.a.	1.3213	0.07	6
BB-2	7.711	n.a.	1.2320	285.23	13
EAet OX3	7.132	0.032	1.1857	16.79	14
EAet OX3,RX	7.110	0.023	1.1833	13.88	18
EAe OX3,RX	7.122	0.040	1.1828	46.05	17
EAe OX3,RX,10%	7.089	0.032	1.1813	60.46	17
EAe GOX,RX	7.114	0.032	1.1821	47.54	17
EAe GOX,*	7.108	0.032	1.1815	46.57	17
EAe GOX,RX,10%	7.079	0.025	1.1803	47.62	18
EAe GOX,*,10%	7.080	0.026	1.1802	48.26	18
EAe GOX,*,100%	7.174	0.058	1.1868	37.45	15
EAebb GOX,*,10%,1%	7.079	0.027	1.1796	69.49	19
EAebb GOX,*,10%,10%	7.076	0.019	1.1791	267.18	21

We compare the greedy heuristic FFFWS, the B&B algorithms BB-1 and BB-2, the EA variants, and the commercial optimizer XOPTS of Albat & Wirsam GmbH. The new algorithms were implemented in C++ and tested on a Pentium 4 PC with 2.8 GHz. Results of XOPTS were provided by Soglatec GmbH and do not respect wagon scheduling. Thus, some of XOPTS' cutting patterns are actually infeasible for our specific 2BP-WS problem. Nevertheless, results from XOPTS provide an important basis for comparison, since 2BP-WS has not been addressed before.

In BB-2, due to memory and time restrictions, the enumeration for a sheet was aborted when the subproblem list exceeded 3 000 000 entries; the best cutting pattern so far was kept.

Strategy parameters of the EAs were determined by extensive preliminary experiments. A population size of 1 000 has been used, and each run was terminated after 50 000 iterations without improvement of the best solution. The number of mutations applied to each new candidate solution is chosen as a Poisson-distributed random variable with expected value 2.

Each EA variant has been tested with different combinations of the presented recombination and mutation operators. In some experiments more than one mutation operator was applied with equal probability; for each mutation the operator to be used was chosen at random. For each EA configuration, 30 independent runs per instance have been carried out.

Average results over all instances obtained by selected algorithm variants are listed in Table 1. Shown are average solution qualities $(\overline{F})$, corresponding average standard deviations $(\overline{\sigma})$ for the EAs, solution qualities relative to the continuous lower bounds $\overline{F/\text{CLB}}$, average total running times $\bar{t}$, and the number of instances for which the algorithm yielded the best solution on average.

It can be observed that average solution qualities do in general not differ very much. FFFWS is the fastest, but usually provides poor results. While BB-1 yields on average the worst results, the solutions of BB-2 are significantly better. However BB-2 also suffers from the greedy nature of optimizing each sheet independently; it is slow and highly memory consuming for larger instances; occasionally, the B&B optimization of some sheets had to be prematurely aborted. BB-2 yields sometimes better, but on average worse results than XOPTS.

In general, the EA variants obtained the best solutions and needed only moderate running times. As expected, EAet is significantly faster than EAe due to its smaller search space. EAebb is slower than EAe due to its B&B optimization of stripes. EAebb with GOX, the combination of all three mutation operators, Lamarckian write-back with probability $p_{wb} = 10\%$ and B&B optimization of stripes applied with a probability of $p_{bb} = 10\%$ performed best with respect to solution quality.

Table 2 provides more detailed results on each of the 31 instances for FFFWS, BB-1, BB-2 and a choice of EA configurations. For the EAs, columns F_{avg} list average solution values over 30 runs per instance, and columns F_{min} show solution values of the best runs.

It turns out that not a single EA configuration is consistently best for all tested instances. On average, however, the element representation yielded significantly better results than the element type representation on 9 instances, and in case of the element representation, GOX outperformed OX3. Lamarckian write-back with probability $p_{wb} = 10\%$ generally increases the quality of final solutions. Regarding mutation, the combined application of RX, BX, and GM, proved to be slightly more robust than only applying RX. The B&B optimization of stripes as done in EAebb increases the quality of solutions a bit further. A Wilcoxon rank test indicates a significant improvement between solutions from EAe and EAebb with $p_{bb} = 10\%$ (GOX, $p_{wb} = 10\%$, and combined mutation used in both cases) on an error level of less than 1% for 6 instances.

Two examples for cutting patterns obtained by FFFWS and EAebb for instance 12 are shown in Fig. 3.

8   Conclusion

We considered different heuristic algorithms for a real-world glass cutting problem, in which the loading of produced elements imposes difficult constraints on feasible cutting patterns. Therefore, approaches for 2BP such as XOPTS cannot be used directly. The proposed FFFWS heuristic is fast and well suited as decoder for an order-based EA. For the EA with element representation, GOX and GM were proposed to encourage sequences of elements of the same type in order to better exploit the possibility of stacking. Incorporating B&B in the decoding for occasionally locally optimizing stripes turned out to increase the solution quality in a few cases. Solutions obtained by the EA are highly satisfactory for practice.

Table 2. Detailed results of selected algorithms. Overall best (average) solution values are printed bold. For BB-2, a mark "a" in column F indicates that in one or more runs B&B had to be aborted for one or more sheets.

inst.	groups	m	n	CLB	XOPTS	FFFWS		BBHEU		BBALG			EAet OX3,RX				EAe GOX,*,10%				EAebb GOX,*,10%,10%			
					F	F	t	F	t	F	t		F_{avg}	σ	F_{min}	t_{avg}	F_{avg}	σ	F_{min}	t_{avg}	F_{avg}	σ	F_{min}	t_{avg}
1	3	3	4	0.83	**0.98**	1.37	0.01	**0.98**	0.00	**0.98**	0.01		**0.98**	0.00	0.98	2.93	**0.98**	0.00	0.98	7.87	**0.98**	0.00	0.98	6.89
2	3	3	4	0.90	**0.95**	0.97	0.02	**0.95**	0.00	**0.95**	0.01		**0.95**	0.00	0.95	2.70	**0.95**	0.00	0.95	8.46	**0.95**	0.00	0.95	7.62
3	3	3	5	0.58	**0.74**	0.83	0.01	0.86	0.00	**0.74**	0.01		**0.74**	0.00	0.74	2.40	**0.74**	0.00	0.74	4.19	**0.74**	0.00	0.74	4.88
4	3	3	6	2.17	**2.88**	2.88	0.02	**2.88**	0.00	**2.88**	0.01		**2.88**	0.00	2.88	5.01	**2.88**	0.00	2.88	7.21	**2.88**	0.00	2.88	8.78
5	1	2	7	0.69	**0.80**	0.99	0.02	0.83	0.01	**0.80**	0.01		0.81	0.00	0.81	3.41	**0.80**	0.00	0.80	5.85	**0.80**	0.00	0.80	6.17
6	2	3	7	2.60	**2.87**	2.96	0.01	3.54	0.01	**2.87**	0.01		**2.87**	0.00	2.87	2.97	**2.87**	0.00	2.87	4.79	**2.87**	0.00	2.87	4.43
7	2	7	9	0.34	0.40	0.46	0.01	0.41	0.01	0.46	0.01		**0.38**	0.00	0.38	4.15	**0.38**	0.00	0.38	6.84	**0.38**	0.00	0.38	15.41
8	3	14	11	1.61	1.91	2.48	0.01	2.53	0.01	1.88	8.72		**1.85**	0.01	1.84	5.39	**1.85**	0.01	1.84	9.56	1.86	0.02	1.84	14.23
9	12	12	14	0.76	0.84	0.99	0.03	**0.83**	0.01	0.99	25.21	a	**0.83**	0.00	0.83	4.40	**0.83**	0.00	0.83	7.69	**0.83**	0.00	0.83	34.43
10	3	7	19	2.08	**2.44**	2.47	0.00	2.67	0.01	**2.44**	64.67	a	**2.44**	0.00	2.44	6.25	**2.44**	0.00	2.44	10.21	**2.44**	0.00	2.44	25.87
11	3	5	22	2.50	**5.56**	5.56	0.00	**5.56**	0.00	**5.56**	12.11	a	**5.56**	0.00	5.56	13.64	**5.56**	0.00	5.56	22.57	**5.56**	0.00	5.56	22.14
12	12	12	33	3.50	3.98	4.67	0.00	4.60	0.01	**3.94**	28.22		3.96	0.00	3.96	10.43	3.95	0.01	3.93	19.01	**3.94**	0.01	3.93	30.57
13	1	16	33	3.65	4.70	4.47	0.01	4.66	0.03	4.24	141.53	a	**4.17**	0.00	4.16	7.51	**4.17**	0.01	4.16	13.15	**4.17**	0.01	4.16	58.23
14	2	24	34	6.72	8.77	8.39	0.01	8.39	0.02	**7.66**	4.47		**7.66**	0.02	7.66	12.04	7.69	0.06	7.66	18.68	**7.66**	0.00	7.66	60.19
15	5	9	37	2.80	2.96	3.48	0.01	3.58	0.03	2.96	31.35		2.97	0.00	2.97	5.57	**2.95**	0.00	2.95	11.18	**2.95**	0.00	2.95	50.96
16	9	9	53	7.86	10.56	10.78	0.02	12.56	0.02	11.56	2.43		**10.24**	0.00	10.24	8.73	**10.24**	0.00	10.24	15.51	**10.24**	0.00	10.24	32.07
17	15	23	60	6.88	7.89	8.68	0.01	8.41	0.03	7.88	257.48		7.87	0.08	7.79	21.57	7.89	0.11	7.79	40.12	**7.80**	0.04	7.67	258.81
18	11	11	61	4.40	4.58	5.15	0.03	5.57	0.02	4.88	2689.97	a	4.64	0.01	4.57	8.80	4.60	0.02	4.56	23.92	**4.56**	0.01	4.55	70.26
19	7	9	69	7.63	**8.22**	8.94	0.01	9.54	0.02	8.54	16.66		8.39	0.01	8.39	9.69	8.25	0.03	8.22	28.83	8.27	0.01	8.22	51.44
20	1	1	70	5.56	6.97	8.68	0.02	**6.93**	0.00	**6.93**	0.03		**6.93**	0.00	6.93	8.24	**6.93**	0.00	6.93	26.48	**6.93**	0.00	6.93	43.35
21	29	31	72	5.28	**5.68**	5.96	0.02	6.66	0.14	5.89	209.92	a	**5.68**	0.03	5.63	29.18	5.75	0.05	5.67	45.60	5.73	0.05	5.64	830.37
22	31	31	72	5.28	5.65	5.96	0.00	6.66	0.15	5.89	215.79	a	**5.67**	0.02	5.64	27.41	5.75	0.05	5.67	43.25	5.75	0.04	5.68	1493.95
23	10	15	77	15.42	18.84	26.49	0.01	20.53	0.03	19.28	621.53	a	**18.53**	0.00	18.53	12.58	**18.53**	0.00	18.53	30.77	18.55	0.01	18.53	82.57
24	4	4	78	1.78	1.99	2.09	0.02	2.00	0.02	2.00	44.96	a	**1.98**	0.00	1.98	7.97	**1.98**	0.00	1.98	19.78	**1.98**	0.00	1.98	106.90
25	2	2	91	4.98	**5.40**	5.89	0.02	5.53	0.01	**5.40**	2.88		5.66	0.00	5.66	6.57	5.49	0.00	5.49	28.73	5.48	0.01	5.45	33.32
26	15	78	140	23.90	30.87	32.69	0.02	38.70	0.95	36.56	127.96		30.09	0.47	29.50	78.44	29.82	0.35	29.19	131.76	**29.80**	0.30	29.45	3484.31
27	3	4	147	8.00	**8.89**	9.16	0.01	9.60	0.03	**8.89**	382.09	a	9.15	0.00	9.15	12.81	8.92	0.04	8.89	69.13	**8.89**	0.00	8.89	78.42
28	11	45	149	22.85	**28.62**	32.62	0.02	37.62	0.38	36.62	979.34	a	28.63	0.02	28.62	29.53	28.63	0.02	28.62	66.87	28.63	0.02	28.62	315.66
29	6	13	151	3.53	3.78	4.00	0.02	4.19	0.05	3.75	398.56	a	**3.70**	0.01	3.68	17.03	3.71	0.01	3.68	63.18	**3.70**	0.01	3.66	178.21
30	8	15	151	9.68	**10.43**	10.93	0.03	12.59	0.03	11.34	387.25	a	10.62	0.02	10.58	23.00	10.49	0.04	10.44	103.95	10.49	0.02	10.47	354.93
31	7	7	651	21.54	23.88	23.79	0.05	26.32	0.14	24.28	2173.64	a	23.56	0.00	23.56	39.88	**23.44**	0.01	23.44	600.92	23.56	0.01	23.53	796.30

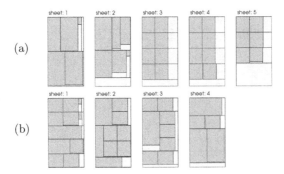

Fig. 3. Cutting patterns for instance 12 obtained by FFFWS (a) and EAebb with GOX, combined RX/BX/GM mutation, $p_{bb} = 10\%$, and $p_{wb} = 10\%$.

References

1. J. O. Berkey and P. Y. Wang. Two-dimensional finite bin packing algorithms. *Journal of the Operational Research Society*, 38:423–429, 1987.
2. E. Coffman, Jr., M. Garey, and D. Johnson. Approximation algorithms for bin packing: A survey. In D. Hochbaum, editor, *Approximation Algorithms for NP-Hard Problems*, pages 46–93. PWS Publishing, Boston, 1996.
3. F. Corno, P. Prinetto, M. Rebaudengo, M. S. Reorda, and S. Bisotto. Optimizing Area Loss in Flat Glass Cutting. In *IEEE Int. Conf. on Genetic Algorithms in Engineering Systems: Innovations and Applications*, pages 450–455, 1997.
4. L. Davis, editor. *A Handbook Of Genetic Algorithms*. Int. Thomson Computer Press, 1991.
5. H. Dyckhoff, G. Scheithauer, and J. Terno. Cutting and packing: An annotated bibliography. In M. Dell'Amico, F. Maffioli, and S. Martello, editors, *Annotated Bibliographies in Combinatorial Optimization*, pages 393–412. Wiley, 1997.
6. E. Hopper. *Two-Dimensional Packing Utilising Evolutionary Algorithms and Other Meta-Heuristic Methods*. PhD thesis, University of Wales, Cardiff, U.K., 2000.
7. S.-M. Hwang, C.-Y. Kao, and J.-T. Horng. On solving rectangle bin packing problems using GAs. In *Proceedings of the 1994 IEEE Int. Conf. on Systems, Man, and Cybernetics*, pages 1583–1590, 1997.
8. B. Kröger. Guillotineable bin packing: A genetic approach. *European Journal of Operational Research*, 84:545–661, 1995.
9. A. Lodi, S. Martello, and M. Monaci. Two-dimensional packing problems: A survey. *European Journal of Operational Research*, 141:241–252, 2002.
10. A. Lodi, S. Martello, and D. Vigo. Heuristic and metaheuristic approaches for a class of two-dimensional bin packing problems. *INFORMS Journal on Computing*, 11:345–357, 1999.
11. S. Martello and D. Vigo. Exact solutions of the two-dimensional finite bin packing problem. *Management Science*, 44:388–399, 1998.
12. M. Monaci. *Algorithms for Packing and Scheduling Problems*. PhD thesis, University of Bologna, Italy, 2002.
13. R. Morabito and M. N. Arenales. Staged and constrained two-dimensional guillotine cutting problems: An AND/OR-graph approach. *European Journal of Operational Research*, 94(3):548–560, 1996.
14. J. Puchinger. Verfahren zur Lösung eines Glasverschnittproblems. Master's thesis, Vienna University of Technology, Austria, 2003.

Designing Reliable Communication Networks with a Genetic Algorithm Using a Repair Heuristic

Dirk Reichelt[1], Franz Rothlauf[2], and Peter Gmilkowsky[1]

[1] Institute of Information Systems, Ilmenau Technical University
Helmholtzplatz 3, P.O. Box 100565, 98684 Ilmenau, Germany,
{Dirk.Reichelt@tu-ilmenau.de,Peter.Gmilkowsky}@tu-ilmenau.de
[2] University of Mannheim, Department of Information Systems I
Schloss, 68131 Mannheim, Germany
rothlauf@uni-mannheim.de

Abstract. This paper investigates GA approaches for solving the reliable communication network design problem. For solving this problem a network with minimum cost must be found that satisfies a given network reliability constraint. To consider the additional reliability constraint different approaches are possible. We show that existing approaches using penalty functions can result in invalid solutions and are therefore not appropriate for solving this problem. To overcome these problems we present a repair heuristic, which is based on the number of spanning trees in a network. This heuristic always generates a valid solution, which when compared to a greedy cheapest repair heuristic shows that the new approach finds better solutions with less computational effort.

1 Introduction

The optimal design of reliable communication and transportation networks is important in many application fields such as gas pipelines, communication networks, and electricity distribution. When designing reliable communication networks there is a trade-off between the necessary investments in the network and the quality of service provided to the network users. An important service measurement is the all-terminal reliability of the network which is defined as the probability that the network is still connected even if some nodes or links fail [1]. In the reliable communication network problem [2] communication links must be chosen such that the network costs are minimized given a network reliability constraint. Both the network design problem and the calculation of the network reliability, have been proven to be NP-hard [3,4]. Genetic Algorithms (GA) have shown promising results when applied to this problem [5,6,7].

In this paper we investigate existing GA approaches for the design of reliable communication networks and propose a heuristic that repairs each candidate solution with respect to the number of spanning trees in the graph. In contrast to other approaches, which only indirectly measure reliability, the number of

J. Gottlieb and G.R. Raidl (Eds.): EvoCOP 2004, LNCS 3004, pp. 177–187, 2004.
© Springer-Verlag Berlin Heidelberg 2004

spanning trees in a graph is a more accurate measurement for the all-terminal reliability of a network [8]. We present empirical results that show that the proposed heuristic outperforms a standard greedy repair heuristic by finding better solutions and using a lower computational effort.

In the following section we give a short problem description. Section 3 investigates different approaches to consider reliability constraints in GA design. In section 4 we discuss the deficites of existing approaches and propose an approach based on the number of spanning trees. Experimental results and a comparison to a simple greedy heuristic are presented in section 5.2. The paper ends with concluding remarks.

2   Problem Definition

For the reliable communication network design problem (RCND) [2], a network topology with minimal cost must be found that satisfies a given reliability constraint. This problem has been proven as NP-hard [3], and several GA-approaches have been proposed for this problem. [9] introduced a branch and bound algorithm minimizing network costs under a reliability constraint. Later, Dengiz et al. proposed a GA [5] using a penalty function to incorporate the reliability constraint directly into the fitness function, as well as a simulated annealing approach [10]. [11] extended this work and developed a parallel GA for larger problem instances. [12] presented a GA with multiple reliability constraints. [6] did not incorporate the reliablity contraint into the fitness function but used a problem specific representation and adapted GA operators.

The RCND problem can be defined as follows: an undirected graph is denoted as $G = (V, E)$, $n = |V|$ denotes the number of nodes, and $m = |E|$ denotes the number of edges of G. It is assumed that the location of each node is given a priori and all nodes are perfect reliable. The degree $d(i)$ of a node i is the number of edges that are connected to node i. For each possible edge $e_{ij} \in E$ the corresponding costs c_{ij} and reliability r_{ij} are known. The probability $1 - r_{ij}$ that the edge e_{ij} fails is statistically independent. A graph G is n-connected if there are at least n edge-disjoint paths between all pairs of nodes i and j. The objective function of the problem is:

$$C(G) = \sum_{i=1}^{n} \sum_{j=i+1}^{n-1} c_{ij} x_{ij} \rightarrow min \tag{1}$$

$$\text{subject to: } R(G) \geq R_0,$$

where $C(G)$ is the total cost of the network G and c_{ij} is the costs for an edge connecting node i and j. The variable $x_{ij} \in (0, 1)$ indicates whether edge $e_{ij} \in E$. $R(G)$ is the all-terminal reliability that is the probability that the network G is still connected (even if some of the edges $e_{ij} \in E$ fail). The calculation of the all-terminal reliability has been proven as NP-hard [4]. Exact algorithms for calculating the all-terminal realiability for networks with a low number of nodes have been proposed by [1,13]. For larger networks, monte carlo-based estimations of the all-terminal reliability [14] are more appropriate. It was shown in [8] that the number of spanning trees in G is an appropriate measurement for the all-terminal reliability (a network G is still connected (and reliable) as long as there is at least one spanning tree in G).

3   Considering Reliability Constraints in Genetic Algorithms

Standard GAs are not able to handle additional problem constraints. Therefore, much research has been focused on how to consider constraints in GA design. It can be distinguished between two different approaches on how to deal with constraints [15,16]. Firstly, indirect constraint handling techniques consider constraints by modifications of the fitness functions. Violations of constraints lead to a lower fitness value (penalty) of the candidate solution. Secondly, direct constraint handling techniques modify the structure of the GA. In principle, there are four different approaches:

- Leave invalid solutions in the population.
- Eliminate infeasible solutions from the population.
- Prevent infeasible solutions by problem-specific representations and operators.
- Repair infeasible candidate solutions.

There are some approaches that have no explicit mechanisms to consider additional constraints but to some extent accept invalid solutions [17]. They hope that the best solution at the end of the run is valid. However, such approaches can only be used if the number of invalid solutions is low.

Other GA approaches eliminate invalid solutions that are generated during a GA run. This approach is only possible if the number of invalid solutions is low. Furthermore, there is the problem that the removal of infeasible candidates solutions may take valuable genetic material from the population that might produce high-quality offspring after recombination and mutation [16,18].

After discussing in general the first two simple direct constraint handling techniques, we focus in the next subsection on the remaining direct, as well as indirect, approaches in the context of the RCND problem.

3.1   Penalty Functions

The indirect constraint handling by penalties as suggested in [19] incorporates a constraint into the fitness function. This transforms a constraint optimization problem to an unconstrained problem by adding penalties for constraint violations to the fitness value of a solution. When using penalties, infeasible solutions remain in the population and their genetic material can be used. Using penalties requires a well-designed penalty function that does not generate new local optima, or let global optima become suboptimal [20].

In the context of the RCND problem, [5] proposed a fitness function with a quadratic penalty term. The objective function from equation 1 becomes:

$$C'(G) = \sum_{i=1}^{n} \sum_{j=i+1}^{n-1} c_{ij} x_{ij} + \delta * (c_{max}(R(G) - R_0))^2$$

$$\delta = \begin{cases} 0, & \text{if } R(G) \geq R_0 \\ 1, & \text{if } R(G) < R_0 \end{cases} \tag{2}$$

$$c_{max} = \max_{e_{ij} \in E} (c_{ij})$$

This problem formulation uses a quadratic penalty term. Additionally, [5] used a repair heuristic which ensures that the degree of all nodes is larger than one $(d(i) \geq 2, \forall i \in V)$.

Table 1. Results of GAs using the penalty approach from equation 2

test instance	r_{ij}	R_0	$C(G_{opt})$	$C'(G'_{opt})$	$R(G'_{opt})$
8 nodes - network 1	0.9	0.90	208	194.222	0.899198
	0.9	0.95	247	223.073	0.949009
8 nodes - network 3	0.9	0.9	211	211	0.902212
	0.9	0.95	245	233.067	0.949948

To check if this penalty approach results in correct solutions, we implemented a GA with the fitness function from equation 2. For the experiments we used a steady state GA with a binary representation, uniform crossover without mutation, and an exact reliability evaluation based on [1].

Table 1 shows the results of our experiments using the proposed penalty function for selected 8 nodes test problems using $r_{ij} = 0.9$ and $R_0 = 0.9$ resp. $R_0 = 0.95$. The two test instances (network 1 and network 3) are taken from [5]. We show the total cost of the correct optimal solution $C(G_{opt})$ published in [5], where $R(G_{opt}) \geq R_0$, the lowest found cost $C'(G'_{opt})$ of the network G'_{opt} according to equation 2, and the corresponding all-terminal realiability $R(G'_{opt})$. It can be seen that using equation 2 can result in solutions G'_{opt} that have lower cost $C'(G'_{opt}) < C'(G_{opt})$, but violate the reliability constraint $(R(G'_{opt}) < R_0)$. Only for one instance (network 3, $r_{ij} = 0.9$, and $R_0 = 0.9$), could a valid solution be found. For the other problems the penalty from equation 2 is too low to ensure a valid solution. As a result the total fitness for an infeasible network G'_{opt} can be smaller than the fitness for the cheapest feasible network G_{opt}. These examples show that an unfavorable design of penalty functions may cause the solutions with lowest fitness to be infeasible. Summarizing the results, the proposed penalty function from [5] does not work in an effective way and can result in invalid solutions.

3.2 Problem-Specific Representations and Operators

Most standard GAs use binary representations and standard operators like n-point or uniform crossover. When applying such standard operators to valid solutions encoded with a standard representation, the resulting offspring can be invalid. This situation can be avoided by using either problem-specific representations, or operators that consider the constraint at hand.

   We want to give two examples for network problems where the optimal solution should be a tree. Trees are a special variant of fully connected graphs G where $|E| = |V| - 1$. The use of the problem-specific Prüfer number representation [21] allows us to consider the constraint that valid solutions are trees. Another possibility to consider this constraint is using direct representations and problem-specific operators (e.g. [22]). The problem-specific operators ensure that only valid solutions (trees) can be created.

   [6] presented problem-specific crossover and mutation operators for the RCND problem. The crossover operator randomly exchanges one link between two parents. If the offspring does not satisfy the reliability constraint, an additional heuristic is applied such that the order of each node is greater than one $(d(i) \geq 2, \forall i \in V)$. The mutation operator searches for two rings in the graph that share only one common node. To reduce the cost $C(G)$ of the network, it merges the two rings to one single ring. If the parent is a network, where $d(i) \geq 2 \ \forall i \in V$, the offspring is also a network with $d(i) \geq 2$.

4 Repair Heuristics

This section shows some deficits of exisiting repair heuristics for the RCND problem, and proposes a new heuristic based on counting spanning trees.

4.1 Deficits of Existing Approaches

When using standard GA operators, problem-specific heuristics can be used to repair invalid solutions violating constraints. A repair heuristic changes candidate solutions such that they become feasible [16]. Two different repair strategies can be distinguished: The Lamarkian approach replaces the parental individual by the offspring. The Baldwinian approach leaves the individual untouched but only its fitness is replaced by the fitness of the repaired solution.

   [6] and [5] introduced greedy repair heuristics for the RCND problem. An individual is repaired such that all nodes have at least the degree two $(d(i) \geq 2, \forall i \in V)$. The repair strategies used the degree of the nodes as a measurement of the all-terminal reliability of the network. However, a network with $d(i) \geq 2, \forall i \in V$, is not always 2-connected as can be seen for example in Figure 1. If the edge $e_{3,4}$ fails, the network is separated into two unconnected network components. Although all

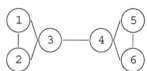

Fig. 1. Example tree with $d(i) \geq 2, \forall i \in V$

nodes in the original network have degree larger than one, it is already disconnected if only one link fails. This example illustrates that the repair procedures
proposed in [5,6] only use a weak reliability measure. To design networks based
on the all-terminal reliability, more accurate measurements of reliability are necessary.

4.2   Spanning Tree Counting Repair Heuristic

In the previous paragraphs we have discussed the deficits of existing approaches
solving the RCND problem. Therefore, inspired by the reliability improvement
procedure proposed by [8], we introduce a GA using a spanning tree counting
(STC) repair heuristic. As the exact calculation of the all-terminal reliability
causes high computation effort, [8] use the number of spanning trees in the graph
G as a measurement of all-terminal reliability. It was shown that the number
of spanning trees in the graph is a good approximation for the all-terminal
reliability.

The basic idea of the STC repair heuristic is to add these edges to the graph
that maximize the reliability (number of spanning trees in the network) with
minimal additional costs. Consequently, the STC repair heuristic calculates for
the cheapest edges that are not in the network G, the possible increase of spanning trees if these edges are added:

1. Sort all links $e_{ij} \notin E$ according to the corresponding edge costs c_{ij}. $i = 0$.
2. Insert the i-cheapest edge e_{ij} temporarily into G and calculate the ratio
   $s_{ij} = c_{ij}/$increase in number of spanning trees in G. $i = i + 1$.
3. If $i < t$ continue with step 2.
4. Add edge $e_{ij} \notin E$ with highest corresponding s_{ij} to G.
5. Calculate $R(G)$.
6. If $R(G) < R_0$, then continue with step 4.

In step two and three the heuristic calculates for the cheapest edges, the ratio
between the cost of the edge, and the increase in the number of spanning trees.
The increase in the number of spanning trees can be calculated with low computational effort by a simple update procedure [8]. The number of edges that are
investigated is limited by t, where $t < n(n-1)/2 - |E|$. In step four to six edges
with highest improve ratio s_{ij} are iteratively added to G until the reliability
constraint is fullfilled.

5   Experiments

5.1   Experimental Design

For our experiments we use a steady state GA with a binary representation of
length $l = n(n-1)/2$. The existence of an edge in G is encoded by 1, its absence
by 0. The GA uses one-point crossover and bit-flipping mutation. The initial
population consists of randomly created 2-connected graphs. The initialization

routine firstly creates a random spanning tree and then randomly adds links until the graph is 2-connected.

As the effort for calculating network reliability is high, we used several techniques to speed up reliability evaluation. As a first step in calculating the all-terminal reliability of a network we determined an upper bound $R_{up}(G)$ for the reliability of a network G using a method proposed by [23]. If $R_{up}(G) < R_0$ the network can not fullfill the reliability constraint and it is not necessary to calculate the reliability exactly. Only if $R_{up}(G) \geq R_0$, we calculate $R(G)$ exactly using a method proposed by [1]. When using this method, we already get a measurement of the all-terminal realiability during the run. We stop the exact calculation as soon as the reliability constraint R_0 is satisfied. Finally, to avoid calculating the reliability of networks G that have been evaluated previously, we store the reliability of all graphs using a hash table. For all new individuals the hash is searched if the network reliability has already been calculated.

If the network reliability of a network is too low ($R(G) < R_0$) the STC heuristic repairs the network and adjusts the chromosome according to the new network. For comparison we have implemented an additional greedy cheapest repair procedure. This heuristic adds the cheapest edge to the network until the graph satisfies the constraint. Unlike the STC heuristic, it does not consider reliability (increase in number of spanning trees) when choosing edges. After constructing a valid solution by a repair heuristic the fitness of the individual is calculated according to equation (1). This approach ensures that we have only valid solutions.

In our experiments we use a steady state GA with 50% replacement, a crossover probability $p_{cross} = 0.9$, a mutation probability $p_{mut} = 0.01$, a population size of 100, an edge reliability $r_{ij} = 0.9, \forall e_{ij} \in E$, and a reliability constraints $R_0 = 0.9$ and $R_0 = 0.95$. The GA stops after 250 generations or convergence. For each test instance we performed ten independent runs.

5.2   Results

Both heuristics have been tested with network problems (8, 10 and 11 nodes) taken from [5], and a new test problem for the 15 largest German cities. We compare the quality of the solutions and the number of repair operations that are necessary for finding high quality solutions. As after each repair operation a reliability check has to be performed, and the reliability checks are computationally demanding, the number of repair operations impacts the running time of the GA. Unfortunately it is not possible to compare our approach directly to the results from [5], because they penalize invalid solutions using the objective function from equation 2, and repair invalid solutions with regard to the degree of the nodes ($d(i) \geq 2, \forall i \in N$). The penalty approach can not be used for a comparison as it can result in invalid solutions (compare section 3.1).

Figure 2 compares the results for the STC heuristic and the greedy cheapest repair heuristic. The plots show the fitness of the best solution and the number of repair operations over the number of generations for the 8 nodes (Figure 2(a)), 10 nodes (Figure 2(b)), and the new 15 nodes (Figure 2(c)) problem. All values

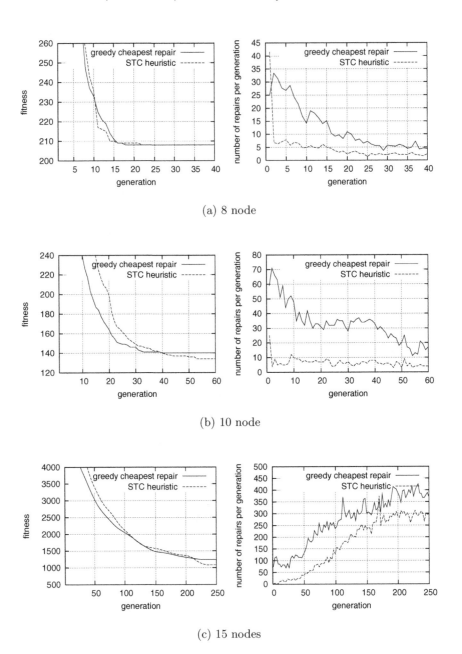

(a) 8 node

(b) 10 node

(c) 15 nodes

Fig. 2. Fitness of the best solution (left) and number of repair operations (right) over the number of generations. The plots show that the STC heuristic finds better solutions and needs less repair operations

are averaged over ten runs. The plots show that the STC heuristic converges more slowly towards high-quality solutions, but always finds better solutions at the end of the run. The plots for the number of repairs show that the STC repair heuristic needs significantly less repairs. Therefore, the STC heuristic is for the 15 nodes problem much faster in comparison to the greedy cheapest repair heuristic (compare also Table 2). This advantage in running time can not be observed for the 8 nodes and 10 nodes problem. The used all-terminal reliability calculation needs for both test problem a low computational effort, therefore the additional repairs have only little impact on the GA running time.

Table 2. Comparison of STC and greedy cheapest repair heuristic

nodes	test problem	R_0	optimum	method	best found	repairs	t_{conv}
8	1	0.9	208	STC	208	158	4 sec
				greedy	208	300	4 sec
8	1	0.9	203	STC	203	658	4 sec
				greedy	203	1548	4 sec
8	2	0.95	194	STC	194.9	595	4 sec
				greedy	207.5	1749	4 sec
8	3	0.9	211	STC	211.6	526	4 sec
				greedy	211.8	2416	4 sec
8	3	0.95	197	STC	198.7	583	4 sec
				greedy	199.1	2789	4 sec
8	4	0.9	291	STC	294.4	700	4 sec
				greedy	296.5	2071	4 sec
8	4	0.95	276	STC	282.1	618	4 sec
				greedy	283.2	2956	4 sec
10	1	0.9	131	STC	134	428	28 sec
				greedy	140.5	1884	29 sec
10	2	0.9	154	STC	155.8	585	27 sec
				greedy	166.7	2488	27 sec
10	2	0.95	136	STC	137.6	1379	26 sec
				greedy	145.5	2479	28 sec
10	3	0.9	267	STC	268	1339	26 sec
				greedy	269.1	1987	26 sec
10	3	0.95	236	STC	243.2	1650	20 sec
				greedy	244.7	2891	20 sec
11		0.9	246	STC	246.7	747	49 sec
				greedy	248.4	1691	50 sec
15		0.9	1006.9	STC	1086.6	44025	9120 sec
				greedy	1217.24	68358	44230 sec

Table 2 summarizes the results for the two heuristics and shows the optimal solution, the average best solution found at the end of a GA run, the average number of repair operations that are necessary to find the best solution, and the average running time t_{conv}. The optimal solutions for the 8, 10 and 11 nodes problem have been published in [5]. The optimal solution for the 15 nodes test problem is the best ever found solution from a GA using the STC heuristic. As before a GA using the STC heuristic finds better solutions with lower computational effort. The additional effort of the STC heuristic for calculating the potential increase in the number of spanning trees in G (compare section 4.2) is low, as it can be computed as the determinant of the node degree matrix [24].

6   Conclusions

This paper investigates existing GA approaches for the reliable communication network design (RCND) problem and proposes a heuristic repair approach based on the number of spanning trees in a network. The analysis of existing approaches for solving the RCND problem reveals some deficits. The penalty approach from [5] can result in invalid solutions and the greedy repair heuristics introduced by [6] and [5] repair invalid solutions according to the degree of the nodes and do not consider the all-terminal reliability of the network.

Therefore, we present a spanning tree counting (STC) repair heuristic that can be combined with standard GAs. This heuristic considers the number of spanning trees in a graph as a more meaningful reliability measure when repairing invalid solutions. The empirical results show that the STC heuristic outperforms a greedy cheapest repair heuristic that considers only the cost of links. The STC heuristic allows only valid solutions and finds in comparison to the greedy cheapest repair heuristic better solution using less computational effort.

References

1. Yubin Chen, Jiandong Li, and Jiamo Chen. A new algorithm for network probabilistic connectivity. In *Proceedings of the IEEE military communication conference*, Piscataway, NJ, 1999. IEEE Service Center.
2. Robert R. Boorstyn and Howard Frank. Large-scale network topological optimization. *IEEE Transactions on Reliability*, 25:29–37, 1977.
3. M. R. Garey and D. S. Johnson. *Computers and Intractibility: A Guide to the Theory of NP-Completeness*. W. H. Freeman and Company, San Fransisco, 1979.
4. Li Ying. Analysis method of survivability of probabilistic networks. *Military Communication Technology Magazine*, 48, 1993.
5. B. Dengiz, F. Altiparmak, and A. E. Smith. Local search genetic algorithm for optimal design of reliable networks. *IEEE Trans. on Evolutionary Computation*, 1(3):179–188, 1997.
6. Sheng-Tzong Cheng. Topological optimization of a reliable communication network. *IEEE Transactions on Reliability*, 47(3):225–233, 1998.
7. Baoding Liu and K. Iwamura. Topological optimization model for communication network with multiple reliability goals. *Computer and Mathematics with Applications*, 39:59–69, 2000.
8. N. Fard and Tae-Han Lee. Spanning tree approach in all-terminal network reliability expansion. *computer communications*, 24:1348–1353, 2001.
9. Rong-Hong Jan, Fung-Jen Hwang, and Sheng-Tzong Chen. Topological optimization of a communication network subject to a reliability constraint. *IEEE Transactions on Reliability*, 42(1):63–70, 93.

10. B. Dengiz and C. Alabas. A simulated annealing algorithm for design of computer communication networks. In *Proceedings of World Multiconference on Systemics, Cybernetics and Informatics, SCI 2001*, volume 5, 2001.

11. Benjamin Baran and Fabian Laufer. Topological optimization of reliable networks using a-teams. In *Proceedings of World Multiconference on Systemics, Cybernetics and Informatics - SCI '99 and ISAS '99*, volume 5, 1999.

12. B.Liu and K. Iwamura. Topological optimization models for communication network with multiple reliability goals. *Computers and Mathematics with Applications*, 39:59–69, 2000.

13. K.K. Aggarwal and Suresg Rai. Reliability evaluation in computer-communication networks. *IEEE Transactions on Reliability*, 30(1):32–35, 1981.

14. E. Manzi E., M. Labbe, G. Latouche, and F. Maffioli. Fishman's sampling plan for computing network reliability. *IEEE Transactions on Reliability*, 50(1):41–46, 2001.

15. Zbigniew Michalewicz. *Genetic Algorithms + Data Structures = Evolution Programs*. Springer-Verlag, Berlin, 3 edition, 1996.

16. B.G.W. Craenen, A.E. Eiben, and E.Marchiori. How to handle constraints with evolutionary algortihms. In Lance Chambers, editor, *Practical Handbook Of Genetic Algorithms: Applications*, pages 341–361. Chapman & Hall/CRC, 2000.

17. L. Davis, D. Orvosh, A. Cox, and Y. Qiu. A genetic algorithm for survivable network design. In S. Forrest, editor, *Proceedings of the Fifth International Conference on Genetic Algorithms*, pages 408–415, San Mateo, CA, 1993. Morgan Kaufmann.

18. Steven Orla Kimbrough, Ming Lu, David Harlan Wood, and D.J. Wu. Exploring a two-population genetic algorithm. In Erick Cantu-Paz et al, editor, *Proceedings of the Genetic and Evolutionary Computation Conference 2003*, pages 1148–1159, Berlin, 2003. Springer-Verlag.

19. D. E. Goldberg. *Genetic algorithms in search, optimization, and machine learning*. Addison-Wesley, Reading, MA, 1989.

20. Jens Gottlieb. *Evolutionary Algorithms for Constrained Optimization Problems*. PhD thesis, Technische Universität Clausthal, Institut für Informatik, Clausthal, Germany, 1999.

21. H. Prüfer. Neuer Beweis eines Satzes über Permutationen. *Archiv für Mathematik und Physik*, 27:742–744, 1918.

22. Günther R. Raidl and Bryant A. Julstrom. Edge-sets: An effective evolutionary coding of spanning trees. *IEEE Transactions on Evolutionary Computation*, 7(3):225–239, 2003.

23. A. Konak and A. Smith. An improved general upperbound for all-terminal network reliability. Technical report, University of Pittsburgh, 1998.

24. C.J. Colbourn. *The Combinatorics of Network Reliability*. Oxford University Press, 1987.

Improving Vehicle Routing Using a Customer Waiting Time Colony

Samer Sa'adah, Peter Ross, and Ben Paechter

School Of Computing
Napier University
10 Colinton Road
Edinburgh, Scotland
EH10 5DT
S.Sa'Adah@napier.ac.uk

Abstract. In the vehicle routing problem with time windows (VRPTW), there are two main objectives. The primary objective is to reduce the number of vehicles, the secondary one is to minimise the total distance travelled by all vehicles. This paper describes some experiments with multiple ant colony systems, in particular a Triple Ant Colony System TACS, in which one colony (VMIN) tries to minimise the number of vehicles, one (DMIN) tries to minimise the total distance and a third (CWTsMAX) tries to *maximise* customer waiting time. The inclusion of this third colony improves the results very significantly, compared to not using it and to a range of other options. Experiments are conducted on Solomon's 56 benchmark problems. The results are comparable to those obtained by other state-of-the-art approaches.

1 Introduction

The vehicle routing problem with time windows is an NP-hard problem in which several vehicles are to be used to deliver quantities of goods to a number of customers. Customer i expects a quantity q_i of goods to be delivered no earlier than time r_i (the 'ready time') and the delivery must be complete by time d_i (the 'due date'), and the delivery takes time s_i (the 'service time'). Each vehicle starts from the one depot and may carry goods for delivery to several customers but the total must not exceed the vehicle's maximum capacity M. If a vehicle arrives at a customer before the customer is ready, the vehicle waits. If a vehicle arrives after the ready time of a customer, in this case the customer has waited. The distance from customer i to customer j is t_{ij}; in benchmark problems this is taken to be the Euclidean distance. Vehicles travel at a speed of 1, so it takes time t_{ij} to cover distance t_{ij}. All the vehicles must return to the depot by the finishing time F.

The problem is to minimise the number of vehicles used to serve all the customers, and also to minimise the total distance travelled by those vehicles. These two may be antagonistic; in many papers there is a preference to push down the number of vehicles necessary. The aim of this paper is to discover

J. Gottlieb and G.R. Raidl (Eds.): EvoCOP 2004, LNCS 3004, pp. 188–198, 2004.

the effects of using different types of routing construction and improvement heuristics on the decision making of the ants and to study several different types of ant colony systems in improving the quality of the solutions. We started by trying to replicate a well-known two-ant-colony system [1] that uses one colony to try to reduce vehicle usage and another to reduce total distance, but have not yet been able to recreate its results. Presumably we are still missing some important aspect of the implementation. Despite this, one of the interesting results reported in this paper is that the inclusion of a third colony that tries to m axim ise customer waiting time does produce a very significant improvement in results.

Section 2 of this paper provides a brief research review. Section 3 describes the various multiple-colony systems we have considered. Solomon's benchmark problems are explained briefly in Section 4. Computational results and comparisons are described in Section 5, and then the paper ends up with some conclusions and further work in Section 6.

2   Background Review

The VRPTW problem has been explored using exact and approximation methods over the years, and has been proven to be an NP hard problem in the strong sense as mentioned in [2] [3]. Exact dynamic programming techniques and column generation and Lagrange relaxation techniques were the first methods to be used to solve the 56 benchmark problems that were proposed by Solomon [4] in 1987 and have been widely studied ever since. Many researchers have used such exact techniques to solve a small set of the Solomon's 56 test problems to optimality, but interest quickly moved to heuristic methods. Solomon in 1987 was among the first researchers to use build routing heuristics (like Savings, Time-Oriented NN, I1, I2, I3 and Time-Oriented Sweep) for solving the VRPTW method [4]. Chiang and Russel in 1993 used hybrid heuristics for the VRPTW [5]. Then, Thangiah et al in 1994 used hybrid approaches based on genetic algorithms, simulating annealing and tabu searches to improve the quality of solutions of the Solomon's 56 problems [6].

Rochat and Taillard in 1995 used a probabilistic diversification and intensification technique in local search for vehicle routing [7]. In 1996, Dorigo et al in [8] and Gambardella et al in [9] used a colony system of cooperating ants each to solve symmetric and asymmetric TSPs. Taillard et al in 1997 used a tabu search that employs an adaptive memory with different neighbourhood structures [10]. Shaw in 1998 used constraint programming and local search methods to solve VRPTWs [11]. In 1999, Kilby et al [12] used a Guided Local Search approach. On the other hand, Gambardella et al in 1999 used a multiple colony of ants in [1] to tackle the VRPTW problem. Also, Liu and Shen proposed in [13] a two-stage metaheuristic based on a new neighbourhood structure (that relates routes and nodes).

In 2000, Bräysy et al in [14] presented a new evolutionary algorithm based on a hybridisation of a genetic algorithm and an evolutionary algorithm consisting of several local search and route construction heuristics. Then, Gehring and Homberger 2001 used in [15] a parallel two-phase metaheuristic that combines a (μ, λ)-evolution strategy and a subsequently executed tabu search. Reimann et al 2002 [16] used the well-known Savings algorithm within the framework of an ant system to tackle the vehicle routing problem. Ellabib et al in [3] presented an experimental study of a simple ant colony system in which the author has tested different initial solution techniques and different visibility (desirability) functions. Bräysy [17] presented a new deterministic metaheuristic based on a modification of a variable neighbourhood search for solving the VRPTW. Berger et al 2003 introduced in [18] a route-directed hybrid genetic approach to evolve two populations pursuing different objectives. One population minimizes total travelled distance while the other minimizes temporal constraint violation. And recently Bräysy et al in [19] used a Threshold Accepting metaheuristic (a post-processor technique) to improve upon results obtained from a parallel genetic algorithm and a multi-start local search. This achieves what are generally the best-known results. For further background information the interested reader should consult Bräysy's papers [17] and [19].

3   Multiple Ant Colony Systems

This section describes ingredients of the ant colony systems we study; ant colony systems studied by other researchers may vary in some of the details.

In, say, a two-colony ant system one colony (VMIN) tries to minimise the number of vehicles and the other (DMIN) tries to minimise the total distance travelled, and the system that controls the colonies keeps track of the global best solution found so far.

Each customer has a 'candidate list' of possible next customers. These candidate lists are constructed before the whole process begins, and contain the twenty (say) nearest customers, ordered (say) by closeness. Restricting attention to the candidate lists when searching for the next customer reduces the search space usefully.

Each colony (VMIN or DMIN) has an initialisation step and a cycle step. In the initialisation step, the colony creates a solution with a number of vehicles allocated to it using some convenient initialisation technique (see below). This initial solution could be feasible or infeasible. Then, the pheromone memory is initialised with $\tau_o = 1/(n.J_{\Psi}^h)$, where n refers to the number of visited nodes (depot or vehicle nodes + customer nodes) and J_{Ψ}^h represents the total of travelled distances.

In the cycle step, ten ants start building a solution, each sequentially. In the case of the VMIN colony, the solution of an ant with maximum visited customers is always preferred, even if not all customers are visited. The best solution known to VMIN, of maximum visited customers will be updated whenever one is found. Now, if the new solution of maximum visited customers is feasible (all customers were visited), then the best global solution will be updated and the system and the DMIN colony will be notified about that. Afterwards, the VMIN colony uses the following global update rules in Equations (1) and (2) to increase and update the pheromone trails of the visited edges in two solutions (the solution with maximum visited customers Ψ^{VMIN} and the global best solution Ψ^{gb} that has been found so far.). Here, it means that the pheromone memory has been double updated by two different global update rules. Also, the VMIN colony uses the global update rule $(\tau_{ij} = (1 - \rho)\tau_{ij})$ to decrease the pheromone trails of the unvisited edges in both solutions. Behold that the pheromone trail of the whole visited (or unvisited) edge is updated rather than the pheromone trail of one side of the edge. τ_{ij} refers to the pheromone value on the edge e_{ij} and ρ is the evaporation parameter. J_Ψ^{VMIN} represents the total of travelled distance done by the vehicles of the solution with maximum visited customers. J_Ψ^{gb} refers to the total of travelled distance done by the vehicles of the best global solution.

$$\tau_{ij} = (1 - \rho)\tau_{ij} + \rho/J_\Psi^{VMIN} \qquad \forall(i,j) \in \Psi^{VMIN} \qquad (1)$$

$$\tau_{ij} = (1 - \rho)\tau_{ij} + \rho/J_\Psi^{gb} \qquad \forall(i,j) \in \Psi^{gb} \qquad (2)$$

In the case of the DMIN colony, the solution with smallest total travelled distances is preferred. If the best-selected solution is better than the best global solution in terms of travelled distances, then the best-selected solution will be taken as the new best global solution and the system and the VMIN colony will be notified about that. Later in this colony, the pheromone trails of the visited edges of the best-global solution will be increased using the global update rule in Equation (2). On the other hand, the pheromone trails of the unvisited edges of the best global solution will be decreased using the global update rule $(\tau_{ij} = (1-\rho)\tau_{ij})$. The cycle step of each colony (VMIN or DMIN) will be notified to stop searching through its ants, if a set amount of CPU time has elapsed. If the CPU limit has not been reached and a global improvement in terms of the number of vehicles has been found, both colonies are killed and two new ones are created and the process restarts.

3.1   What the Ants Do

Each ant is essentially looking for a complete solution involving a given number of vehicles. Rather than keeping track of separate vehicles, the problem is reformulated by introducing several 'ghost' depots co-located with the original depot, so that there are as many depots as vehicles; and now an ant is looking for a single closed tour, and the operation of the ant colony discourages an ant from visiting two 'ghost' depots unnecessarily.

There are several major steps in what an ant does. At the first step, an ant selects a random depot to start from. In the second step, the ant builds a solution by selecting edges in the following way. Let b_i be the beginning time of service at customer i, let T_{ij} be the time at which service can begin at customer j, if the ant comes from customer i, and let V_{ij} be the available time between the end of servicing i and the due date of customer j. Let U_j be the number of times that customer j has not yet been visited by the ants of a colony, in VMIN, or 0 in DMIN. Calculate η_{ij} as follows:

$$T_{ij} = \max(b_i + s_i + t_{ij}, r_j) \tag{3}$$

$$V_{ij} = d_j - (b_i + s_i) \tag{4}$$

$$C_{ij} = \max(1.0, T_{ij}V_{ij} - U_j) \tag{5}$$

$$\eta_{ij} = 1/C_{ij} \tag{6}$$

The ant, when at customer i, decides probabilistically whether it is exploiting (with probability p_E, a parameter of the system) or exploring. In exploit mode, it chooses to go to the unvisited customer j on the candidate list of customer i that maximises $\tau_{ij}[\eta_{ij}]^\beta$ (where β is another parameter of the system); in explore mode, it chooses probabilistically among unvisited customers j, where each such customer has a probability p_{ij} of being chosen:

$$p_{ij} = \frac{\tau_{ij}.[\eta_{ij}]^\beta}{\sum_{l \in N_i^k} \tau_{il}.[\eta_{il}]^\beta} \tag{7}$$

Once an edge is chosen, the local pheromone-updating rule in Equation (8) is applied to decrease the pheromone trail in that selected edge.

$$\tau_{ij} = (1 - \rho).\tau_{ij} + \rho.\tau_o \tag{8}$$

After implementing the first two steps to build a solution, an insertion procedure is applied to the solution of an ant to insert all unvisited customers, if this solution is still infeasible. So, the remainder of unvisited customers are ordered in descending order according to their demand quantities. Each unvisited customer will be tried for insertion next to the nearest visited node on the candidate list, either just after or just before it, whichever minimises distance if both options are feasible. If insertion next to the nearest unvisited customer is infeasible, the next nearest is tried and so on.

Finally, in the DMIN colony, ants try to further improve the solutions they have built, using a local search called TriplesMovesLS that has some similarities with the CROSS exchange local search [1]. Given a solution, the TriplesMovesLS local search works through each vehicle's route in turn, and for each customer on the route it tries three possible 'moves':

1. Swap the visited node with the nearest node on the candidate list.
2. Relocate the visited node after the nearest node on the candidate list.
3. Relocate the nearest node on the candidate list after the visited node.

If all three moves are not feasible, then the visited node is tried with its second nearest on its candidate list and so on. If a 'move' improves the solution it is accepted, and the local search continues.

3.2 Specific Multiple Colony Systems

We consider two specific systems, DACS and TACS. In DACS, a double ant colony system, the candidate list for customer i is sorted according to how soon a delivery could be started at j if j were to be the next node visited, and there are just two colonies, VMIN and DMIN. Initial solutions are built on the basis of distance: start from a depot, visit the nearest visitable (non-ghost-depot) customer, continue until a vehicle route has been built and either the vehicle capacity or the finishing-time constraint applies.

In TACS, a third colony is included which tries to maximise total customer waiting time, that is, ensure that deliveries happen close to the due times on average. We have tried other variants in which we instead use a third colony that tries to minimise vehicle waiting time (that is, waiting for customers to become ready) or try to minimise rather than maximise total customer waiting time, but they are not quite so effective. We also tried a quadruple ant colony system, QACS, which used both a colony that tried to minimise total customer waiting time and another that tried to maximise the same thing. A comparison between DACS, TACS and QACS on Solomon's R1 problem set is shown in Table 1. As is customary in VRPTW research, averages for the whole problem set are reported, the customers are located at the same fixed positions for every member of a set, but customer-time windows vary from wide to narrow between members of the set. NV refers to the number of vehicles on average over thirty runs; TD refers to total distance on average. QACS improves NV very slightly but at slight cost in TD. T-tests indicate that the reduction in the number of vehicles, by both TACS and QACS when compared with DACS, is significant at the 99.9% level. In what follows, we do not consider QACS further, but it is still a counter-intuitive result that including two directly antagonistic colonies should still produce very good results!

Table 1. Comparison between DACS, TACS and QACS

	DACS		TACS		QACS	
Problem Name	NV	TD	NV	TD	NV	TD
R1	12.88	1220.92	12.73	1241.05	12.70	1243.06

3.3 Synergetic Effects of Artificial Ants

It may be asked, is the pheromone updating important or is local search really doing the work? In a simple experiment to show the synergetic effects of the artificial ants, the systems (DACS and TACS) have been tested with (and without) pheromone global and local updates. This experiment has been tried on Solomon's problem set R1, see below. The averaged results of thirty runs of the systems in Table 2 show that using the pheromone global and local updates is very significant (at the 99.9% level, according to t-tests) and without such updates the performance of both systems are really very bad in terms of the number of vehicles and the total of travelled distances.

Table 2. Experiment of the synergetic effects of the artificial ants

	DACS		TACS	
R1	NV	TD	NV	TD
With Pheromone	12.88	1220.92	12.73	1241.05
Without Pheromone	13.12	1372.18	13.15	1386.19

4   Solomon's Benchmark Problems

Solomon's benchmark problems [4] consist of six sets of problems. Within a set, the customers are always at the same locations. Sets R1 and R2 have randomly located customers, at the same places in both sets; in R1, vehicles have small capacity and in R2 they have large capacity, and within a set the time windows of the customers vary. Sets C1 and C2 have clustered customers, and differ in the same way as R1 and R2. Sets RC1 and RC2 have a mix of clustered and randomly placed customers. Furthermore, the sets (R1, C1 and RC1) have tight time windows, whereas the sets (R2, C2 and RC2) have wide time windows. Because of the way in which the sets of benchmarks have been constructed, it is common in VRPTW research to report the average number of vehicles required and average total distance travelled, averaged over all the members of a set.

5   Computational Results and Comparisons

In Tables 3 and 4, the results and the performance of our two main systems (DACS and TACS) are compared with the published results mentioned in the literature. Both systems have been coded in Java and run on Pentium IV with 2.0 GHz speed and 512 MB RAM. Table 3 presents the results on the problem sets R1, C1, and RC1, Table 4 those on R2, C2 and RC2. Experiments are made through running each problem instance for three runs. Each run has been on for a fixed amount of CPU time in seconds (100, 300, 400, 600, 1200 or 1800). The computational times of the different approaches in Tables 3 and 4 cannot be directly compared, because of hardware and software differences. For each problem instance of a problem set, the results of the best solutions of three runs have been averaged. Then, the averaged results of all problem instances in a problem set have averaged again and reported. In addition, the results of the best solution of three runs on each problem instance in a problem set have been gathered and the results from all problem instances of a problem

Table 3. Published results on the problem sets R1, C1 and RC1.

	R1			C1			RC1		
	NV	TD	TIME	NV	TD	TIME	NV	TD	TIME
MACS-VRPTW [1]	12.55	1214.80	100	10.00	828.40	100	12.46	1395.47	100
	12.45	1212.95	300	10.00	828.38	300	12.13	1389.15	300
	12.38	1213.35	600	10.00	828.38	600	12.08	1380.38	600
	12.38	1211.64	1200	10.00	828.38	1200	11.96	1385.65	1200
	12.38	1210.83	1800	10.00	828.38	1800	11.92	1388.13	1800
RT [7]	12.83	1208.43	450	10.00	832.59	540	12.75	1381.33	430
	12.58	1202.00	1300	10.00	829.01	1600	12.50	1368.03	1300
	12.58	1197.42	2700	10.00	828.45	3200	12.33	1269.48	2600
SW [11]	12.45	1198.37	900				12.05	1363.67	900
	12.35	1201.47	1800				12.00	1363.68	1800
	12.33	1201.79	3600				11.95	1364.17	3600
KPS [12]	12.67	1200.33	2900	10.00	830.75	2900	12.12	1388.15	2900
CW [20]	12.50	1241.89	1382	10.00	834.05	649	12.38	1408.87	723
TB [10]	12.64	1233.88	2296	10.00	830.41	2926	12.08	1404.59	1877
	12.39	1230.48	6887	10.00	828.59	7315	12.00	1387.01	5632
	12.33	1220.35	13774	10.00	828.45	14630	11.90	1381.31	11264
DACS	13.44	1289.42	100	10.00	860.55	100	13.21	1454.84	100
	13.25	1245.12	300	10.00	855.92	300	12.88	1422.21	300
	13.31	1247.99	400	10.00	844.25	400	12.83	1427.10	400
	13.17	1244.77	600	10.00	860.31	600	12.71	1412.59	600
	13.14	1231.12	1200	10.00	847.71	1200	12.58	1414.30	1200
	13.00	1225.67	1800	10.00	835.91	1800	12.75	1404.20	1800
TACS	12.64	1241.22	1200	10.00	831.44	1200	12.29	1425.16	1200
RVNS(1) [17]	12.00	1229.48	---	10.00	828.38	---	11.50	1394.26	---
RVNS(2) [17]	11.92	1222.12	4950	10.00	828.38	4950	11.50	1389.58	4950
BBB [18]	12.17	1251.40	1800	10.00	828.50	1800	11.88	1414.86	1800
LS [13]	12.25	1253.68	2495	10.00	841.33	1459	12.00	1416.11	1868
HGA+TA [19]	12.17	1208.57	2094	10.00	828.38	2094	11.75	1372.93	2094
LS+TA [19]	12.00	1220.20	156	10.00	828.38	156	11.50	1398.76	156
HM4C [15]	12.00	1217.57	810	10.00	828.63	810	11.50	1395.13	810

set have been averaged and reported in Table 5. All our three systems (DACS, TACS and QACS) have used the following parameters: 10 ants for each cycle of a colony, $p_E = 0.9$, $\beta = 1$ and $\rho = 0.1$. It can been seen from the results and the performance of our systems in Tables 3, 4 and 5 that DACS, although it is essentially a reconstruction of Gambardella's MACS-VRPTW, does not perform as well as MACS-VRPTW (and therefore we are still missing some key component). But TACS, which is DACS with an extra colony, does produce much improved results.

6   Conclusions and Further Work

A Triple Ant Colony System (TACS) has been introduced to solve the VRPTW problem. The model is applied to the well-known Solomon data sets, and the performance is evaluated for each data set based on the average number of vehicles and total of travelled distances. The results show that the performance

Table 4. Published results on the problem sets R2, C2 and RC2.

	R2			C2			RC2		
	NV	TD	TIME	NV	TD	TIME	NV	TD	TIME
MACS-VRPTW [1]	3.05	971.97	100	3.00	593.19	100	3.38	1191.87	100
	3.00	969.09	300	3.00	592.97	300	3.33	1168.34	300
	3.00	965.37	600	3.00	592.89	600	3.33	1163.08	600
	3.00	962.07	1200	3.00	592.04	1200	3.33	1153.63	1200
	3.00	960.31	1800	3.00	591.85	1800	3.33	1149.28	1800
RT [7]	3.18	999.63	1600	3.00	595.38	1200	3.62	1207.37	1300
	3.09	969.29	4900	3.00	590.32	3600	3.62	1155.47	3900
	3.09	954.36	9800	3.00	590.32	7200	3.62	1139.79	7800
SW [11]									
KPS [12]	3.00	966.56	2900	3.00	592.29	2900	3.38	1133.42	2900
CW [20]	2.91	995.39	1332	3.00	591.78	292	3.38	1139.70	946
TB [10]	3.00	1046.56	3372	3.00	592.75	3275	3.38	1248.34	1933
	3.00	1029.65	10116	3.00	591.14	8187	3.38	1220.28	5798
	3.00	1013.35	20232	3.00	590.91	16375	3.38	1198.63	11596
DACS	3.27	1094.37	100	3.29	668.60	100	3.83	1330.21	100
	3.18	1063.77	300	3.08	610.94	300	3.67	1273.33	300
	3.18	1038.20	400	3.08	600.10	400	3.71	1246.69	400
	3.18	1020.19	600	3.00	593.24	600	3.71	1199.88	600
	3.15	994.12	1200	3.00	592.53	1200	3.58	1173.25	1200
	3.18	983.86	1800	3.00	591.92	1800	3.54	1157.52	1800
TACS	3.03	966.08	1200	3.00	590.42	1200	3.38	1165.12	1200
RVNS(1) [17]	2.73	989.62	---	3.00	590.30	---	3.25	1141.07	---
RVNS(2) [17]	2.73	975.12	4950	3.00	589.86	4950	3.25	1128.38	4950
BBB [18]	2.73	1056.59	1800	3.00	590.06	1800	3.25	1258.15	1800
LS [13]	2.82	1022.08	403	3.00	591.03	224	3.25	1230.31	430
HGA+TA [19]	2.73	971.44	2094	3.00	589.86	2094	3.25	1154.04	2094
LS+TA [19]	2.73	970.38	156	3.00	589.86	156	3.25	1139.37	156
HM4C [15]	2.73	961.29	810	3.00	590.33	810	3.25	1139.37	810

Table 5. Average of the best solutions computed by different VRPTW algorithms.

	R1		C1		RC1		R2		C2		RC2	
	NV	TD	NV	TD	NV	TD	NV	TD	NV	TD	NV	TD
MACS-VRPTW [1]	12.00	1217.73	10.00	828.38	11.63	1382.42	2.73	967.75	3.00	589.86	3.25	1129.19
RT [7]	12.25	1208.50	10.00	828.38	11.88	1377.39	2.91	961.72	3.00	589.86	3.38	1119.59
TB [10]	12.17	1209.35	10.00	828.38	11.50	1389.22	2.82	980.27	3.00	589.86	3.38	1117.44
CR [5]	12.42	1289.95	10.00	885.86	12.38	1455.82	2.91	1135.14	3.00	658.88	3.38	1361.14
PB [21]	12.58	1296.80	10.00	838.01	12.13	1446.20	3.00	1117.70	3.00	589.93	3.38	1360.57
TH [6]	12.33	1238.00	10.00	832.00	12.00	1284.00	3.00	1005.00	3.00	650.00	3.38	1229.00
DACS	12.75	1220.90	10.00	829.63	12.38	1415.45	3.18	967.41	3.00	590.28	3.50	1140.58
TACS	12.58	1224.40	10.00	828.38	12.13	1404.99	3.00	953.90	3.00	590.28	3.38	1137.40
RVNS(1) [17]	12.00	1229.48	10.00	828.38	11.50	1394.26	2.73	989.62	3.00	590.30	3.25	1141.07
RVNS(2) [17]	11.92	1222.12	10.00	828.38	11.50	1389.58	2.73	975.12	3.00	589.86	3.25	1128.38
BBB [18]	11.92	1221.10	10.00	828.48	11.50	1389.89	2.73	975.43	3.00	589.93	3.25	1159.37
LS [13]	12.17	1249.57	10.00	830.06	11.88	1412.87	2.82	1016.58	3.00	591.03	3.25	1204.87
LS+TA [19]	12.00	1214.69	10.00	828.38	11.50	1389.20	2.73	960.44	3.00	589.86	3.25	1124.14
HGA+TA [19]	11.92	1216.58	10.00	828.38	11.50	1387.66	2.73	966.05	3.00	589.86	3.25	1143.12
LS+HGA+TA [19]	11.92	1215.62	10.00	828.38	11.50	1385.17	2.73	956.35	3.00	589.86	3.25	1123.84
HM4 [15]	12.08	1208.14	10.00	828.50	11.50	1389.65	2.73	965.46	3.00	589.88	3.25	1126.22
HM4C [15]	12.00	1217.57	10.00	828.63	11.50	1395.13	2.73	961.29	3.00	590.33	3.25	1139.37

of the system improves, normally in terms both of the number of vehicles and total distance travelled, when a colony that tries to maximise customer waiting time is added to the colonies VMIN and DMIN.

In comparison with the other metaheuristics, the results and the performance of the TACS system is getting closer to the best-known results and performance. However, the TACS system is not producing the best results published so far, and we have been unable to replicate published results even by re-implementing published ideas. Our further work will be to look at how to improve the performance and the results of the TACS system through looking at different kinds of route elimination and customer ejection techniques. In addition, different kinds of visibility functions for the ants will be investigated and explored.

References

1. Gambardella, L.M., Taillard, E., Agazzi, G.: MACS-VRPTW: A Multiple Ant Colony System for Vehicle Routing Problems with Time Windows. In D. Corne, M. Dorigo and F. Glover, editors, New Ideas in Optimization. McGraw-Hill, UK, (1999) pp. 63–76. Also available as Technical Report IDSIA-06-99, IDSIA, Lugano, Switzerland.

2. Larsen, J.: Parallelization of the vehicle routing problem time windows. Ph.D. thesis, Institute of Mathematical Modelling, Technical University of Denmark, Lyngby, Denmark (1999)

3. Ellabib, I., Basir, O.A., Calamai, P.: An Experimental Study of a Simple Ant Colony System for the Vehicle Routing Problem with Time Windows. M. Dorigo et al. (Eds.): ANTS 2002, LNCS 2463, (2002) pp. 53–64. ©Springer-Verlag Berlin Heidelberg.

4. Solomon, M.M.: Algorithms for the Vehicle Routing and Scheduling Problem with Time Window Constraints. Operations Research 35(2) (1987) pp. 254–265.

5. Chiang, W.C., Russel, R.: Hybrid Heuristics for the Vehicle Routing Problem with Time Windows. Hybrid Heuristics for the Vehicle Routing Problem with Time WindowsWorking Paper, Department of Quantitative Methods, University of Tulsa, OK, USA. (1993)

6. Thangiah, S., Osman, I., Sun, T.: Hybrid Genetic Algorithm Simulated Annealing and Tabu Search Methods for Vehicle Routing Problem with Time Windows. Technical Report 27, Computer Science Department, Slippery Rock University. (1994)

7. Rochat, Y., Taillard, E.D.: Probabilistic Diversification and Intensification in Local Search for Vehicle Routing. Journal of Heuristics 1 (1995) pp. 147–167.

8. Dorigo, M., Maniezzo, V., Colorni, A.: The Ant System: Optimization by a Colony of Cooperating Agents. IEEE Transactions on Systems, Man, and Cybernetics-Part B, 26(1) (1996) pp. 29–41.

9. Gambardella, L.M., Dorigo, M.: Solving Symmetric and Asymmetric TSPs by Ant Colonies. ICEC96, Proceedings of the IEEE Conference on Evolutionary Computation, Nagoya, Japan, May 20-22. (1996)

10. Taillard, E.D., Badeau, P., Gendreau, M., Guertin, F., Potvin, J.Y.: A tabu search heuristic for the vehicle routing problem with soft time windows. Transportation science 31 (1997) pp. 170–186.

11. Shaw, P.: Using Constraint Programming and Local Search Methods to Solve Vehicle Routing Problems. Proceedings of the Fourth International Conference on Principles and Practice of Constraint Programming (CP '98), M. Maher and J.-F. Puget (eds.), Springer-Verlag, (1998) pp. 417–431.
12. Kilby, P., Prosser, P., Shaw, P.: Guided Local Search for the Vehicle Routing Problems With Time Windows. In Meta-heuristics: Advances and Trends in Local Search for Optimisation, S. Voss, S. Martello, I. H. Osman and C. Roucairol (eds.), Kluwer Academic Publishers, Boston, (1999) pp. 473–486.
13. Liu, F.H., Shen, S.Y.: A Route-Neighborhood-based Metaheuristic for Vehicle Routing Problem with Time Windows. European Journal of Operational Research 118, (1999) pp. 485–504.
14. Bräysy, O., Berger, J., Barkaoui, M.: A new hybrid evolutionary algorithm for the vehicle routing problem with time windows. Presented at the Route 2000-Workshop, Skodsborg, Denmark (2000)
15. Gehring, H., Homberger, J.: Parallelization of a Two-Phase Metaheuristic for Routing Problems with Time Windows. Asia-Pacific Journal of Operational Research 18, (2001) pp. 35–47.
16. Reimann, M., Doerner, K., Doerner, K.: A Savings based Ant System for the Vehicle Routing Problem. Proceedings 4th International Conference on Genetic and Evolutionary Computation (GECCO 2002), New York City, New York, USA, Morgan Kaufmann, San Francisco California, (2002) pp. 1317–1325.
17. Bräysy, O.: A Reactive Variable Neighbourhood Search for the Vehicle Routing Problem with Time Windows. INFORMS Journal on Computing ©2002 INFORMS Vol. 00, No. 0, (2002) pp. 1–22.
18. Berger, J., Barkaoui, M., Bräysy, O.: A Route-directed Hybrid Genetic Approach for the Vehicle Routing Problem with Time Windows. Information Systems and Operational Research 41:2 (2003) pp. 179–194.
19. Bräysy, O., Berger, J., Barkaoui, M., Dullaert, W.: A Threshold Accepting Metaheuristic for the Vehicle Routing Problem with Time Windows. Central European Journal of Operations Research, 4 / (2003) .
20. Cordone, R., Wolfer-Calvo, R.: A heuristic for the vehicle routing problem with time windows. Journal of Heuristics 7 (2001) pp 107–129.
21. Potvin, J.Y., Bengio, S.: The Vehicle Routing Problem with Time Windows - Part II: Genetic Search. INFORMS Journal of Computing 8, (1996) pp. 165–172.

New Benchmark Instances for the QAP and the Experimental Analysis of Algorithms

Thomas Stützle[1] and Susana Fernandes[2]

[1] Darmstadt University of Technology, Computer Science Department,
stuetzle@informatik.tu-darmstadt.de
[2] Universidade do Algarve, Departamento de Matemática, sfer@ualg.pt

Abstract. The quadratic assignment problem arises in a variety of practical settings. It is known to be among the hardest combinatorial problems for exact algorithms. Therefore, a large number of heuristic approaches have been proposed for its solution. In this article we introduce a new, large set of QAP instances that is intended to allow the systematic study of the performance of metaheuristics in dependence of QAP instance characteristics. Additionally, we give computational results with several high performing algorithms known from literature and give exemplary results on the influence of instance characteristics on the performance of these algorithms.

1 Introduction

The QAP can best be described as the problem of assigning a set of objects to a set of locations with given distances between the locations and given flows between the objects. The goal is to place the objects on locations in such a way that the sum of the product between flows and distances is minimal. It is a model of many real world problems arising in hospital layout, keyboard layout and other areas [4,5].

More formally, given n objects and n locations, two $n \times n$ matrices $A = [a_{ij}]$ and $B = [b_{rs}]$, where a_{ij} is the distance between locations i and j and b_{rs} is the flow between objects r and s, the QAP can be stated as

$$\min_{\phi \in \Phi} \sum_{i=1}^{n} \sum_{j=1}^{n} a_{\phi_i \phi_j} b_{ij} \tag{1}$$

where Φ is the set of all permutations of the set of integers $\{1, \ldots, n\}$, and ϕ_i gives the location of object i in the current solution $\phi \in \Phi$.

The QAP is a $\mathcal{NP}$-hard optimization problem [17]. It is considered as one of the hardest optimization problems since the largest instances that can be solved today with exact algorithms are limited to instances of size around 30 [1,13]: the largest, non-trivial instance from QAPLIB [12], a benchmark collection for the QAP, solved to optimality has 36 locations [3]. In practice, the only feasible way to solve large QAP instances is to apply heuristic algorithms which find very high quality solutions in short computation time. Several such algorithms

J. Gottlieb and G.R. Raidl (Eds.): EvoCOP 2004, LNCS 3004, pp. 199–209, 2004.

have been proposed which include algorithms like simulated annealing [6], tabu search [2,18,22], genetic algorithms [10,16,8], GRASP [14], ant algorithms [11, 15,21,20], and scatter search [7].

Previous research results show that the characteristics of the QAP instances strongly influence the relative performance of the various algorithmic techniques for the QAP [23,11,21,19]. Based on the input data, two measures were identified that have significant influence on the relative performance of metaheuristics, the (flow) dominance that corresponds to the variation coefficient, i.e. the standard deviation of the (flow) matrix entries divided by their average value, and the sparsity of the matrix, i.e. the fraction of zero entries [16,21]. Similarly, the performance of metaheuristics for the QAP has been related to search space characteristics like ruggedness and fitness distance correlations [16,21]. However, currently there is still a strong lack of (i) a complete understanding of how exactly the relative performance of different metaheuristics depends on instance characteristics, (ii) knowledge of how input data characteristics of QAP instances and search space features relate one to each other, and (iii) even the relative performance of algorithms is not clear because of the different hardware and implementation details used in the various researches. Our goal is to improve this situation through the introduction of a new, large set of benchmark instances that offers a range of instances with different sparsities and flow dominances and using these instances for a systematic analysis of metaheuristic implementations that are as much as possible based on the same data structures. In this paper, Section 2 describes the set of new benchmark instances and Section 3 exemplifies the analysis of the influence of QAP instance characteristics on the relative performance of various metaheuristic implementations. We conclude in Section 4.

2   Benchmark Instances

2.1   Instances from QAPLIB

It is known that the particular type of a QAP instance has a considerable influence on the performance of heuristic methods [23]. For example, the instances available from QAPLIB have been classified by Taillard into four classes. Despite the fact that QAPLIB comprises about 130 instances, there are several disadvantages associated for experimental analysis. First, many of the QAPLIB instances are too small to pose a real challenge to the best available metaheuristic algorithms for the QAP and only a small number of instances of a size with $n \geq 50$ remain (the largest instance is of size $n = 256$ and only one instance of this size exists).[1] Second, with a small number of instances, one may introduce a strong

---

[1] For example, the high performing iterated local search algorithm from Stützle [19] on all except 18 instances from QAPLIB finds the proven optimal or best known solutions in every single trial of limited length (these results were measured across at least 25 trials of a maximal length of 1200 seconds on a 1.2 GHz Athlon processor); most of these instances are actually "solved" on average within few seconds. However, most QAPLIB instances, including also the small ones of size about 20, are still very challenging for exact algorithms.

bias when solving these instances with metaheuristics due to overfitting effects. Third, the instances in QAPLIB do not vary systematically in their characteristics and therefore their utility is very limited for analyzing the performance of metaheuristics in dependence of instance characteristics.

Other, new instances that were designed to be hard were proposed by Taillard and Drezner [9]. Taillard's instances are available at `http://ina.eivd.ch/collaborateurs/etd/problemes.dir/qap.dir/qap.html` and Drezner's instances are available at `http://business.fullerton.edu/zdrezner/programs.htm`. However, these instances have very particular features and the instance characteristics of the two new classes of instances are rather similar, limiting the scope of experimental studies. Nevertheless, these instances are interesting for studies that examine, for which type of instances metaheuristic instances may fail.

2.2 New Random QAP Instances

Because of the limitations associated with the instances from QAPLIB and those of other sources, we generated a large number of instances with varying characteristics. The set of instances was generated in such a way that (i) instance characteristics are systematically varied and (ii) that it is large enough to allow systematic studies on the dependence of the performance of metaheuristics on instance characteristics. These instances were also used in the experimental evaluation of several metaheuristics in the Metaheuristics Network (see also `www.metaheuristics.net`) and are available at URL `www.intellektik.informatik.tu-darmstadt.de/~tom/qap.html`.

Instance generation. The instance generation is done in such a way that instances with widely different syntactic features concerning the flow dominance values of the matrix entries (the dominance value is the variation coefficient of the matrix entries) and the sparsity of the matrix entries (which measures the percentage of zero matrix entries) can be generated. These two measures strongly influence the characteristics of the QAP instances and are conjectured to also have a strong influence on metaheuristics' performance. As a second important parameter we considered two different ways of generating the distance matrix, which allows to generate distance matrices with different structures. In particular, for the distance matrix we applied two different ways of generating the distance matrix.

Euclidean distances: In a first approach the distance matrix corresponds to the Euclidean distance between n points in the plane. These points are generated in such a way that they fall into several clusters of varying size. The characteristics of the cluster (location, number of points) are generated as follows
  - generate a random number k according to a uniform distribution in $[0, K]$, where K is a parameter.
  - generate a cluster center (c_x, c_y), where c_x and c_y are randomly chosen according to a uniform distribution in $[0, 300]$.

– randomly choose the coordinates of the k cluster points in a square centered around (c_x, c_y):
1. generate a and b randomly according to a uniform distribution in $[-m/2, m/2]$, where m is a parameter,
2. set $x = c_x + a$ and $y = c_y + b$.

These steps are repeated until n points have been generated.

Manhattan distances: In a second approach, the distance matrix corresponds to the pairwise distances between all the nodes on a $A \times B$ grid, where the distance is calculated as the Manhattan distance between the points on a grid.

The reasons for the choice of the different distance matrices are the following. Instances based on Euclidean distances will have most probably only one single optimal solution and by different parameter settings for the generation of the distance matrices, the clustering of locations often encountered in real-life QAP instances can be imitated. Instances with a distance matrix based on the Manhattan distance of points on a grid have, due to symmetries in the matrix, at least four optimal solutions which are at a maximum distance possible from each other[2]. Moreover, the distance matrices show a much lower clustering. Additionally, some real-life instances have distances derived from points on a grid.

Different ways of generating the flow matrix are also considered. The flow matrix for all instances are asymmetric and have a null diagonal. Here, in particular, we generate instances which have a strongly varying flow dominance and sparsity. For the generation of the flow matrix we considered two different cases: (i) the flow entries are randomly generated, (ii) flow entries are structured in the sense that clusters of objects exist which have a significant interaction.

Random flows: The generation of the random flow matrices uses the following parameters:
– sp, with $0 \leq sp < 1$, indicates the sparsity of the flow matrix
– a and b determine the flow values.

Each flow matrix entry is then generated by the following routine

```
double x = random->next(); % a random number in [0,1]
if ( x < sp )
  return 0;
else {
  x = random->next();
  return (unsigned long int) MAX(1,pow((a * x),b));
}
```

where x is a random number uniformly distributed in $[0, 1]$. By varying sp we can generate instances of arbitrary sparsity. The parameters a and b allow to vary the distribution of the matrix entries according to $y = (a \cdot x)^b$.

---

[2] As the distance between two solutions we use here a straightforward extension of the Hamming distance that counts the number of different positions in two permutations corresponding to the number of different assignments.

The instances below are generated with the following combinations of a and b: $(100, 1)$, which corresponds to uniformly distributed matrix entries in $[0, 100]$ (the same as the unstructured instances of one of the instances classes defined by Taillard [23]), $(10, 2), (3.5, 3), (2, 7)$. For fixed value of sp, the instances with larger exponent tend to have higher flow dominance.

Structured flows: In real applications, groups of objects tend to have typically a large interaction, while the interaction among different groups tends to be smaller. Instances with structured flow matrix entries take these properties into account. This is done in an indirect way as follows: We generate n points (objects) randomly according to a uniform distribution in a square of dimension 100×100. In the flow matrix a flow between two objects i and j exists if the distance between the points i and j is below some constant threshold value d: "Close" objects tend to have flow, while for "far" objects the flow is zero. The non-zero flow entries are generated like in the random flow case as $\max\{1, (a \cdot x)^b\}$. Note that the threshold value d has a strong influence on the sparsity (the smaller d, the less objects will exist which exchange flow) and the flow dominance of the flow matrix.

Structured flows plus: In the flow generation of the structured flows, distant objects will not exchange any flow. In a straightforward extension, we generate with a small probability p flows between objects for which the associated points have a distance larger than the threshold d. The non-zero flow entries are generated like in the random flow case as $\max\{1, (a \cdot x)^b\}$.

Benchmark instances. We distinguish among six different classes of instances that differ in the way different variants for the distance matrix and the flow matrix are combined. The class identifiers are

RandomRandom: Random distance matrix and random flows;
RandomStructured: Random distance matrix and structured flows;
RandomStructuredPlus: Random distance matrix and structured flows with connections among clusters of objects;
GridRandom: Grid-based distance matrix and random flows;
GridStructured: Grid-based distance matrix and structured flows;
GridStructuredPlus: Grid-based distance matrix and structured flows with connections among clusters of objects.

For each of the six classes we have randomly generated instances of different sizes with $n \in \{50, 100, 150, 200, 300, 500\}$. For each class a number of instances differing in the above defined parameters for generating the matrix entries were generated resulting in a total of 644 new benchmark instances.

To give an impression of the variety of the instance characteristics within these instances, in Figure 1 we indicate the variation of the flow dominance and the sparsity of the instances across all the 136 instances of size 50. In that plot four different curves are visible. This pattern is due to the influence of the parameters a and b on the generation of the flow matrix. In general, the plot shows that there is a strong, non-linear relationship between the flow dominance and

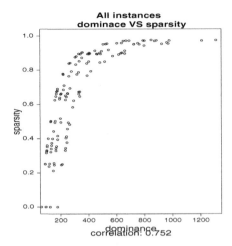

Fig. 1. Plots of the relationship between flow dominance and sparsity of the instances of size 50. On the x-axis is given the flow dominance and on the y-axis is given the sparsity of the instances.

the sparsity. Similarly, we investigated properties like autocorrelation length and the number of improvement steps for reaching a local optimum; these measures give some hint on the ruggedness of the search landscape.

3   Experimental Results

As said, the set of new instances is intended to allow the study of the performance of metaheuristics in dependence of instance characteristics. In this section we give an example of the experimental results and insights that can be obtained using the set of new instances. Here, we limit our analysis to instances of size 50 and give performance results for six different algorithms. In particular, we run robust tabu search (RoTS) [22], a simulated annealing (SA) algorithm [6], $\mathcal{MAX}$–$\mathcal{MIN}$ Ant System ($\mathcal{MMAS}$) [21], two iterated local search (ILS) variants, a population-based one (ES-ILS) and a non-population based one (ILS) [19], and fast ant system (FANT) [24]. All the algorithms were using the parameter settings as proposed by their original authors. The choice of these algorithms was based on the desire to have different types of metaheuristics known to perform well for specific classes of instances. (In an extended version of this article we will include a number of additional algorithms including most of the current well known state-of-the-art algorithms.)

All these algorithms were implemented in C based on the same underlying local search procedure and using the same data structures as much as possible. Each algorithm was run 10 times on each of the instances for 150 seconds on a Pentium III 700 MHz processor under SUSE Linux 7.1. Note that the stopping

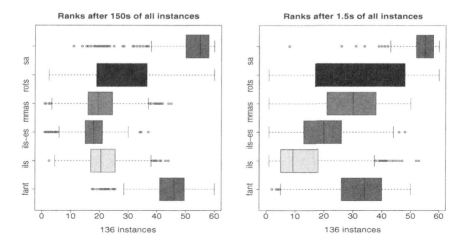

Fig. 2. Boxplots of the ranks for the six metaheuristics measured across all the instances of size 50 (left: stopping criterion 150 seconds; right: stopping criterion 1.5 seconds). The x-axis gives the rank for each single solution found by an algorithm. Note that for each instance a total of 60 trials was run and therefore on each instance the range of the ranks is from 1 to 60.

criterion of 150 seconds was chosen in such a way that RoTS can do about $10\,000n$ iterations on our computer, where n is the instance size.

The experimental analysis was based on ranking the results returned by the algorithms. For every algorithm the best solution achieved in each single trial was saved and then all these values were ranked. Since each of the six algorithms was run ten times, this results overall in 60 possible ranks. Based on the ranks achieved by each algorithm, we used boxplots to visualize the results achieved. In the boxplots, the central box shows the data between the two quartiles, with the median being represented by a line. "Whiskers" extend to the most extreme data point which is no more than 1.5 times the interquartile range from the box. If there are data points further away than the extensions of the whiskers, these data are indicated by small circles.

In Figure 2 we give a summary of the results when ranking across all the 136 instances of size 50 (hence, each box is based on 1360 data points), when stopping the algorithms after 150 seconds (left side) and when stopping the algorithms after 1.5 seconds (right side). The plots suggest that there are significant differences among the algorithms, which is confirmed by using Friedman rank sum tests that rejects the null hypothesis that says that all the results are equal. (We applied the Friedman rank sum test to all the data presented in the following figures; in all cases this test indicates that there exist significant differences in the performance of the algorithms.) In fact, for the largest computation times, ES-ILS appears to be the best performing algorithm followed by $\mathcal{MMAS}$ and ILS. The worst performing algorithms are FANT and SA. The ranking of the algorithms is different, however, if only short computation times are allowed.

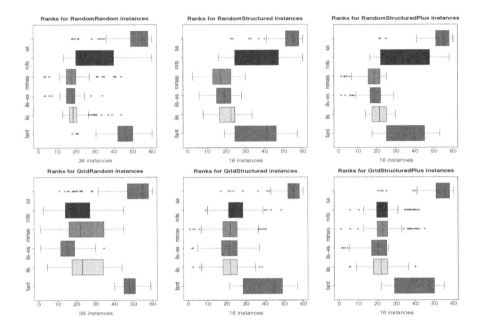

Fig. 3. Boxplots of the ranks for the six metaheuristics measured across the six different instance classes. The x-axis gives the rank for each single solution found by an algorithm. The upper plots are random distance matrices, while the lower plots are for instance classes where the distance matrix is based on a grid.

In fact, when allowing only 1.5 seconds of computation time, ILS gives the best results followed by ES-ILS, $\mathcal{MMAS}$, and RoTS and FANT. The reason that ILS is now much better performing than either ES-ILS or $\mathcal{MMAS}$ is that the latter two algorithms are population-based algorithms and with short computation times, they were only able to run few iterations. Differently, ILS uses only one single solution at each step and therefore it is also quicker in identifying high quality solutions early in the search. It is also noteworthy that FANT is now, relative to the other algorithms, much better performing than for longer trials. In fact, FANT was designed to achieve good solutions quickly [24].

Further analysis was done to investigate the dependence of the relative performance of the algorithms on the type of instances and on features like the flow dominance. In Figure 3 we give results for the six different instance classes for the 150 seconds time limit (see page 203 for a description of the six classes). The results show that the class of instances can have significant influence of the relative performance of the metaheuristics tested here. For example, on the instances with random distance matrix, FANT catches up in performance with RoTS, while it is significantly worse than RoTS on instances with Grid distance matrix. Similarly, $\mathcal{MMAS}$ catches up with ES-ILS on the instances with random distance matrix, while on the instances with grid distance matrix ES-ILS appears to be superior to $\mathcal{MMAS}$.

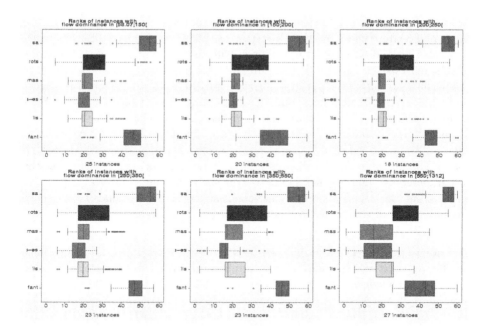

Fig. 4. Boxplots of the ranks for the six metaheuristics measured across the instances classified by the flow dominance. The x-axis gives the rank for each single solution found by an algorithm. Note that for each instance a total of 60 trials was run and therefore on each instance the range of the ranks is from 1 to 60.

In Figure 4 we present the results in dependence of the flow dominance values of the various instances. In particular, we divided the range of observed flow dominance into six intervals from [58; 150],]150; 200],]200; 250],]250; 350],]350; 550],]550; 1312]. For increasing flow dominance, the performance of RoTS is decreasing when compared to ES-ILS, ILS, and $\mathcal{MMAS}$, which can be observed, for example, in the increasing differences concerning the medians obtained by the various algorithms. When comparing ES-ILS to $\mathcal{MMAS}$ it becomes clear that the advantage of ES-ILS over $\mathcal{MMAS}$ is mainly because of the better performance on instance with medium values of the flow dominance in the range from 250 to 550. This observations is true for the largest time limits, while for shorter time limits the advantage of ES-ILS over $\mathcal{MMAS}$ appears to be larger (the corresponding plots are not shown here); this may suggest that $\mathcal{MMAS}$ is taking longer time to converge than ES-ILS.

4   Conclusions

In this paper we introduced a new set of QAP instances that is intended for the systematic analysis of the performance of metaheuristics in dependence of instance characteristics. We exemplified the use of these classes exemplifying the performance differences of six well known metaheuristic implementations using an analysis based on ranking procedures. Currently we are extending the scope of this analysis strongly by (i) including a larger number of metaheuristics into the performance analysis, (ii) applying the algorithms also to the larger instances and to all QAPLIB instances, and (iii) examining the dependence of the algorithms' performance on measures based on search space characteristics (instead of pure syntactic features of the instance date) like fitness-distance correlation, the ruggedness of the search spaces, or the occurrence of plateaus in the search space. The ultimate goal of this research is to deepen our understanding of the performance of different metaheuristics on the QAP and, more in general, to learn about the relation between metaheuristic performance and search space characteristics of the problem under solution.

Acknowledgments. We would like to thank Luis Paquete for his suggestions on this research and his help with R and several scripts to extract the results. Susana acknowledges support from a joint DAAD-ICCTI project. This work was supported by the Metaheuristics Network, a Research Training Network funded by the Improving Human Potential programme of the CEC, grant HPRN-CT-1999-00106. The information provided is the sole responsibility of the authors and does not reflect the Community's opinion. The Community is not responsible for any use that might be made of data appearing in this publication.

References

1. K. M. Anstreicher, N. W. Brixius, J.-P. Goux, and J. Linderoth. Solving large quadratic assignment problems on computational grids. *Mathematical Programming*, 91(3):563–588, 2002.
2. R. Battiti and G. Tecchiolli. The reactive tabu search. *ORSA Journal on Computing*, 6(2):126–140, 1994.
3. N. W. Brixius and K. M. Anstreicher. The Steinberg wiring problem. Technical report, College of Business Administration, University of Iowa, Iowa City, USA, October 2001.
4. R. E. Burkard, E. Çela, P. M. Pardalos, and L. S. Pitsoulis. The quadratic assignment problem. In P. M. Pardalos and D.-Z. Du, editors, *Handbook of Combinatorial Optimization*, volume 2, pages 241–338. Kluwer Academic Publishers, 1998.
5. E. Çela. *The Quadratic Assignment Problem: Theory and Algorithms*. Kluwer Academic Publishers, Dordrecht, The Netherlands, 1998.
6. D. T. Connolly. An improved annealing scheme for the QAP. *European Journal of Operational Research*, 46(1):93–100, 1990.
7. V.-D. Cung, T. Mautor, P. Michelon, and A. Tavares. A scatter search based approach for the quadratic assignment problem. In T. Baeck, Z. Michalewicz, and X. Yao, editors, *Proceedings of the 1997 IEEE International Conference on Evolutionary Computation (ICEC'97)*, pages 165–170. IEEE Press, Piscataway, NJ, USA, 1997.
8. Z. Drezner. A new genetic algorithm for the quadratic assignment problem. *INFORMS Journal on Computing*, 15(3):320–330, 2003.

9. Z. Drezner, P. Hahn, and É. D. Taillard. A study of quadratic assignment problem instances that are difficult for meta-heuristic methods. *Annals of Operations
   Research*, to appear.
10. C. Fleurent and J. A. Ferland. Genetic hybrids for the quadratic assignment
    problem. In P. M. Pardalos and H. Wolkowicz, editors, *Quadratic Assignment
    and Related Problems*, volume 16 of *DIMACS Series on Discrete Mathematics and
    Theoretical Computer Science*, pages 173–187. American Mathematical Society,
    Providence, RI, USA, 1994.
11. L. M. Gambardella, É. D. Taillard, and M. Dorigo. Ant colonies for the quadratic
    assignment problem. *Journal of the Operational Research Society*, 50(2):167–176,
    1999.
12. P. Hahn. QAPLIB - a quadratic assignment problem library.
    http://www.seas.upenn.edu/qaplib/, 2003. Version visited last on 15 September
    2003.
13. P. M. Hahn, W. L. Hightower, T. A. Johnson, M. Guignard-Spielberg, and C. Rou-
    cairol. Tree elaboration strategies in branch and bound algorithms for solving
    the quadratic assignment problem. *Yugoslavian Journal of Operational Research*,
    11(1):41–60, 2001.
14. Y. Li, P. M. Pardalos, and M. G. C. Resende. A greedy randomized adaptive
    search procedure for the quadratic assignment problem. In P. M. Pardalos and
    H. Wolkowicz, editors, *Quadratic Assignment and Related Problems*, volume 16
    of *DIMACS Series on Discrete Mathematics and Theoretical Computer Science*,
    pages 237–261. American Mathematical Society, Providence, RI, USA, 1994.
15. V. Maniezzo. Exact and approximate nondeterministic tree-search procedures for
    the quadratic assignment problem. *INFORMS Journal on Computing*, 11(4):358–
    369, 1999.
16. P. Merz and B. Freisleben. Fitness landscape analysis and memetic algorithms for
    the quadratic assignment problem. *IEEE Transactions on Evolutionary Computa-
    tion*, 4(4):337–352, 2000.
17. S. Sahni and T. Gonzalez. P-complete approximation problems. *Journal of the
    ACM*, 23(3):555–565, 1976.
18. J. Skorin-Kapov. Tabu search applied to the quadratic assignment problem. *ORSA
    Journal on Computing*, 2(1):33–45, 1990.
19. T. Stützle. Iterated local search for the quadratic assignment problem. Technical
    Report AIDA-99-03, FG Intellektik, FB Informatik, TU Darmstadt, 1999.
20. T. Stützle and M. Dorigo. ACO algorithms for the quadratic assignment problem.
    In D. Corne, M. Dorigo, and F. Glover, editors, *New Ideas in Optimization*, pages
    33–50. McGraw Hill, London, UK, 1999.
21. T. Stützle and H. H. Hoos. $\mathcal{MAX}$–$\mathcal{MIN}$ Ant System. *Future Generation Com-
    puter Systems*, 16(8):889–914, 2000.
22. É. D. Taillard. Robust taboo search for the quadratic assignment problem. *Parallel
    Computing*, 17(4–5):443–455, 1991.
23. É. D. Taillard. Comparison of iterative searches for the quadratic assignment
    problem. *Location Science*, 3(2):87–105, 1995.
24. É. D. Taillard. FANT: Fast ant system. Technical Report IDSIA-46-98, IDSIA,
    Lugano, Swiss, 1998.

Improving Edge Recombination through Alternate Inheritance and Greedy Manner

Chuan-Kang Ting

International Graduate School of Dynamic Intelligent Systems
University Paderborn, 33098 Paderborn, Germany
ckting@upb.de

Abstract. Genetic Algorithms (GAs) are well-known heuristic algorithms and have been widely applied to solve combinatorial problems. Edge recombination is one of the famous crossovers designed for GAs to solve combinatorial problems. The essence of edge recombination is to achieve maximal inheritance from parental edges. This paper presents two strategies to improve edge recombination. First, we encourage alternation of parents in edge inheritance. Second, a greedy method is used to handle the failures occurred in edge recombination. A modified edge recombination, called edge recombination with tabu (Edge-T), is proposed according to these two strategies. The traveling salesman problem is used as a benchmark to demonstrate the effectiveness of the proposed method. Experimental results indicate that Edge-T can achieve better performance than the conventional edge recombination Edge-3 in terms of both solution quality and convergence speed.

1 Introduction

Genetic algorithms (GAs) are well-known heuristic algorithms based on the imitation of natural evolution [Hol75]. Their effectiveness in search and optimization problems has been extensively validated. Crossover is one of the most salient features of GAs. It simulates the creation of offspring in the natural world by exchanging and recombining genetic information from two selected parents. This operator is believed to be capable of exploring the problem space effectively. However, in order to enhance GAs' performance, several crossovers have been proposed to deal with specific problems. For example, in terms of combinatorial problems, there exist partially mapped crossover (PMX) [GL85], order crossover (OX) [Dav85], and cycle crossover (CX) [OSH87].

The traveling salesman problem (TSP) is a classic combinatorial problem. Given n cities, the objective of TSP is to find the shortest tour that visits all cities exactly once. This problem has been proven NP-complete; in other words, under the general assumption $P \neq NP$, there is no exact algorithm which can solve the problem in polynomial time. Heuristic algorithms, therefore, are commonly used to deal with this problem.

To solve the TSP with a GA, the crossovers used for order based representations, such as PMX, OX, and CX, are traditionally adopted. The character of these cross

J. Gottlieb and G.R. Raidl (Eds.): EvoCOP 2004, LNCS 3004, pp. 210–219, 2004.

overs is to permute and recombine the order information of parental chromosomes. The adjacency information, however, is crucial to TSP but is not utilized in these operators. Whitley et al. [WSF89] proposed a series of crossovers, called *edge recombination*, to overcome this drawback. Edge recombination utilizes an edge table to record the adjacency of parental cities, and then builds the filial tour with reference to this edge table and specific heuristics. The method (Edge-1) and its modifications (Edge-2,3,4,5,6) have shown their superiority over PMX, OX, and CX in TSP [Sta91, NYY00].

Generally, edge recombination attempts to inherit as many parental edges as possible. Its ability to preserve parental edges has been validated through its higher correlation coefficient between the fitness of parents and offspring [MdWS91]. This ability is believed useful to exploit the fitness landscape and is capable of better performance. However, another adjacency-based crossover, edge assembly crossover (EAX) [NK97, NK99] uses different strategies from such a "maximal inheritance" tactic. First, an *AB-cycle* is defined as a cycle formed by alternating parental sequences. Second, a so-called *E-set* collects a subset of AB-cycles by two methods: random selection and heuristic selection. The intermediate sub-tours are then established on the basis of the E-set. Finally, EAX uses a greedy manner to merge these sub-tours. This procedure attempts to connect two sub-tours with the edges which can achieve the minimal total distance of these two sub-tours. Experimental results show that EAX achieves a breakthrough in the performance of crossover in TSP [Wat98]

Considering EAX's success we propose the following two strategies to enhance edge recombination.

1. Inheritance of edges from alternate parents.
2. Greedy manner in connecting sub-tours.

Instead of maximal inheritance, these two strategies result in more diverse inheritance. The alternation of parental inheritance can not only preserve parental information but also utilize this information in a more explorative way. The original edge recombination attempts to generate highly correlated offspring with their parents. This highly-correlated inheritance, nevertheless, poses a risk of the biased inheritance, which inherits edges merely from one parent and consequently renders the population similar. An alternation of parents can prevent this drawback and maintain population diversity. In addition, the concept of maximal inheritance also affects the failure treatment in edge recombination. The original edge recombination and its modifications all try to recover and extend the tour-building process when failure occurs. However, these deadlock tour segments can be viewed as sub-tours. Our second strategy, therefore, does not prevent failures from introducing foreign edges as the original edge recombination do. Instead, our proposed greedy manner attempts to introduce an *advantageous* foreign edge to connect these sub-tours.

In this paper, a modified edge recombination, called *edge recombination with tabu* (Edge-T), is proposed to validate the effectiveness of these two strategies in edge recombination. First, a tabu list [GK97] is applied to Edge-T in order to restrain inheriting the successive edge from the same parent. Such a restriction is expected to increase the alternate level of parents. Second, Edge-T adopts a greedy method rather than a recovery manner to handle the failure condition. Several TSP instances, ranging from 51 to 442 cities, are used as benchmarks to evaluate the performance of Edge-T.

The rest of this paper is organized as follows. Section 2 gives a brief description of edge recombination. Next, Section 3 describes the algorithm Edge-T and its strategies in a more detail. Section 4 presents experimental results including the influence of Edge-T's components and performance comparisons. Finally, conclusions are drawn in Section 5.

2   Edge Recombination

Edge recombination is a crossover which focuses on the adjacency relation. As illustrated in Figure 1, edge recombination uses an edge table to record parental edges, and then limits the generation of offspring to the edges contained in this table. In other words, the candidates of offspring edges come from parental edges principally.

With reference to the edge table, edge recombination builds a tour according to specific heuristics. In the original edge recombination (Edge-1) [WSF89], the building process intends to select the city with the fewest links, namely, the most *isolating* city. Edge-1 initially generates the edge table by scanning both parents. Afterwards, Edge-1 begins the process of building the filial tour. (1) Select one of the first parental cities as the starting city. (2) Select its next city from the remaining adjacent cities (links) of the current city. We call these links *candidates* for the next city. According to the heuristic's "priority of isolating cites", the candidates with the smallest number of links is chosen. (3) Repeat (2) until the complete tour is built.

City	Links	#Links	City	Links	#Links
A	B,J,H	3	F	E,G,C	3
B	A,C,I,H	4	G	F,H,C,I	4
C	B,D,F,G	4	H	G,I,B,A	4
D	C,E,J	3	I	H,J,G,B	4
E	D,F	2	J	I,A,D	3

Fig. 1. Edge table of edge recombination

Several modifications based on different heuristics have been proposed to improve the original edge recombination. Edge-2 [Sta91], for instance, pays more attention to common edges of both parents than the isolating cites. It marks the common link with a negative (-) sign to indicate its priority over other odd links. Experimental results demonstrate this heuristic can improve the original edge recombination. In addition to the process of building a tour, some modifications focus on handling the failure condition. In edge recombination a failure occurs when the crossover can not find any possible candidate for the next city. The original edge recombination solves this deadlock by randomly choosing the next city from the unselected cities. In this case, the crossover is forced to introduce a *foreign* edge, which means an edge that does not exist in parental edges. Such a foreign edge is believed to be harmful to the performance of crossover. In order to reduce the number of failures, Edge-3 [MW92] reverses the built tour segment when a failure occurs, and then continues the building process from the *live* end instead of the *dead* end. If another failure occurs thereafter, the algorithm will randomly choose the next city as Edge-1 does because there is no *live* end for this built tour segment. This randomly chosen city subsequently acts as a beginning of a new tour segment and is viewed as another *live* end. Edge-4 [DW94], nonetheless, only reverses a partial segment rather than the whole built segment. Furthermore, Edge-5 [NYY00] employs another heuristic of choosing the next city to limit the number of failures. Edge-5 intends to select the edges from the same parent during building tour. The results show this approach can significantly reduce the failure rate at the mature stage. Edge-6 [NYY00] further deals with the failures in Edge-5. When a failure occurs, Edge-6 examines if there is an edge connecting two ends of the built tour segment. If it holds, Edge-6 will seek a city along the built tour to link an unselected city. Then Edge-6 breaks the connection of the sought city and its successive city, and re-connects it with the unselected city. Experimental results indicate Edge-5 and Edge-6 can decrease the failure rate and obtain better performance.

3   Edge Recombination with Tabu (Edge-T)

The *edge recombination with tabu* (Edge-T) is proposed to validate our two strategies in enhancing edge recombination. First, we argue that the alternation of parents in inheritance is capable of better performance. To meet this purpose, Edge-T incorporates a tabu list to prevent inheriting the successive edge from the same parent. Second, we consider a greedy manner in handling failures can improve the performance. Here a built tour segment that encounters a failure is viewed as a sub-tour. Edge-T attempts to introduce the shortest foreign edge to connect these sub-tours.

To incorporate the first strategies with edge recombination, Edge-T uses a priority function to sum up the weights of common edges and isolating level. Furthermore, the factor of tabu restriction is added to this priority function as a penalty. The function used to evaluate the *priority* of an edge from the current city i to its link j is thus defined:

$$P_{ij} = 2m_{ij} + (4 - l_j) - T_{ij} \tag{1}$$

where m_{ij} denotes the number of common edges, l_j denotes the number of remaining links for city j, and T_{ij} is the number of tabu occurrences. Edge-T always chooses the link with the greatest priority value. The first term of the priority function indicates the portion of common edges. In the priority function only the value m_{ij} greater than 1 will be taken into calculation; otherwise this value is set to 0. The second term expresses the portion of isolating level. The number of remaining links l_j ranges from 0 to 4 and is inversely proportional to the isolating level. To meet the *maximization* of the priority function, we subtract the value l_j from its maximal value 4. The last term, T_{ij}, accounts for the penalty factor. A *tabu list* [GK97] is applied to record the prohibitive parents for inheritance. In addition, a corresponding *ancestry list* is necessarily added to each link to trace the parents where the edge inherits from.

As Figure 2 shows, for example, in terms of city A, the link to city B comes from parent 1; the link to city J is a common edge of parent 1 and parent 2. When an edge is selected from the edge table, the ancestry list of this edge turns into the tabu list, which inhibits the building process to choose the next edge from the same parent.

Edge Table:

City	Links	#Links	City	Links	#Links
A	B(1),J(1,2),H(2)	3	F	E(1,2),G(1),C(2)	3
B	A(1),C(1),I(2),H(2)	4	G	F(1),H(1),C(2),I(2)	4
C	B(1),D(1),F(2),G(2)	4	H	G(1),I(1),B(2),A(2)	4
D	C(1),E(1,2),J(2)	3	I	H(1),J(1),G(2),B(2)	4
E	D(1,2),F(1,2)	2	J	I(1),A(1,2),D(2)	3

Tabu List:

1	

Fig. 2. The extended edge table and tabu list of Edge-T

Totally speaking, these three portions – common edges, isolating level, and tabu restriction – constitute the decision criteria for building tours. Following the previous example, if the offspring tour starts with city A, it will have three available links to consider. The tabu list is empty at this time because there is no previous edge. Figure 3(a) shows the priority of its candidates B, J, H is 1, 6, and 1 respectively. Due to the greatest priority value, city J is picked and then its ancestry list turns into the tabu list as shown in Figure 3(b). This "evaluate-and-choose" process will continue until the offspring tour is finished or a failure occurs.

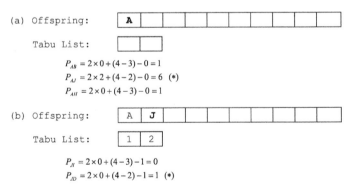

(a) Offspring: | A | | | | | | | | | |

Tabu List: | | |

$$P_{AB} = 2 \times 0 + (4-3) - 0 = 1$$
$$P_{AJ} = 2 \times 2 + (4-2) - 0 = 6 \quad (*)$$
$$P_{AH} = 2 \times 0 + (4-3) - 0 = 1$$

(b) Offspring: | A | J | | | | | | | | |

Tabu List: | 1 | 2 |

$$P_{JI} = 2 \times 0 + (4-3) - 1 = 0$$
$$P_{JD} = 2 \times 0 + (4-2) - 1 = 1 \quad (*)$$

Fig. 3. Example of 10-city tour building in Edge-T

Failures, nevertheless, occur probably in Edge-T as well as in edge recombination. Unlike edge recombination's preventive way, Edge-T adopt a greedy manner to introduce foreign but advantageous edges instead. When a failure occurs, namely no available edge to choose, the built tour segment can be viewed as a *sub-tour*. During the building process, there may be some of such sub-tours. In this case, Edge-T will seek the *shortest* link between two sub-tours. Specifically, Edge-T will search the unselected cities for the shortest link from the end of currently built segment. The complexity of this search process is $O(n)$, where n is the number of cities in TSP. This computation cost is not so expensive but will bring a considerable advantage to the performance of crossover.

4   Experiments

In this paper, a series of experiments are conducted to evaluate the performance of Edge-T. These experiments adopt eight TSP instances [TSPLIB] which are widely used as benchmarks. Table 1 lists these experimental instances and their respective optimal tour length.

The schemes of GA employed in our experiments are steady-state GA, path representation, linear-ranking selection with bias 1.25. Here we set complete crossover ($p_c = 1.0$) and no mutation ($p_m = 0$) to examine the absolute influence of crossover on performance, although the utilization of mutation may result in better solutions. The survival strategy is to delete the worst member of a family, which consists of two parents and their children. A population size of 500, 1000, and 2000 chromosomes is empirically set for problems fewer than 300 cities, lin318, and pcb442 respectively. Each experimental setting includes 30 independent runs where population is initialized randomly.

The experiments first examine the impact of tabu list upon the alternation of parental edges. Second, convergence comparisons are presented to verify the performance of Edge-T in subsection 4.2.

4.1   Influence of Tabu List

Edge-T applies a tabu list to restrain the edge inheritance from the same parent. We examine the effectiveness of this tabu restriction with an inspection upon the composition of a tour. Here only the results on the TSP instance d198 (198 cities) is presented because of the similar results as other instances. Figure 4 shows the proportions of edges from alternate parents, successive parents, and failure treatment. The result indicates that Edge-T has a steadily higher proportion of alternate parents than Edge-3. This outcome validates the effectiveness of tabu restriction in alternating parents. Its corresponding influence on convergent performance will be further checked in the next subsection. In addition, Edge-T has a higher failure proportion than that of Edge-3. This higher failure rate of Edge-T reveals that the failure treatment plays an important role in the performance.

4.2 Performance Comparison

The effectiveness of Edge-T on solution quality and convergence speed is further examined by comparison with Edge-3, Edge-Tx (Edge-T without tabu restriction), and Edge-Ty (Edge-T without tabu restriction and using the failure treatment of Edge-3). Table 1 and Table 3 summarize the experimental results on TSP instances ranging from 51 to 442 cities.

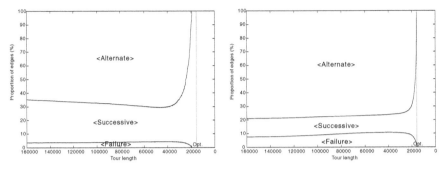

Fig. 4. Proportion of edges from alternate parents, successive parents, and failure treatment of Edge-3 (left) and Edge-T (right).

Table 1 compares the solution quality. Edge-T achieves the best solution quality on average and the smallest standard deviation on all experiments. Edge-Tx, meanwhile, obtains better solution and smaller standard deviation than Edge-3 and Edge-Ty does. However, there is no clear difference between Edge-3 and Edge-Ty. A series of one-tailed t-tests is further performed to validate the significance of superiority in terms of solution quality. Table 2 presents the statistical results of t-test on the experimental data of Table 1. With a confidence level $\alpha = 0.01$, the optimal solutions obtained from Edge-3 and Edge-Ty do not differ significantly on most problems except lin318 and pcb442. It indicates the priority function of Edge-T does not result in significant difference in performance from Edge-3. In addition, the statistical results $p < 0.01$ in the (EdgeTy − EdgeTx) and (Edge3 − EdgeTx) column validate that Edge-Tx can achieve better solution than Edge-Ty and Edge-3. These two outcomes support that the greedy manner can improve the failure treatment of Edge-3. Furthermore, the proposed algorithm Edge-T, achieves even better solution quality than Edge-Tx according to the t-test results in the (EdgeTx − EdgeT) column of Table 2. This preferable outcome confirms the effectiveness of tabu restriction on the advanced improvement of Edge-Tx. As shown in Figure 4, the tabu restriction results in not only a larger-scale parental alternation but also a higher failure rate. The superior performance of Edge-T, therefore, is attributed to a joint effort of these two effects.

Table 1. Comparative tour length of Edge-3 and Edge-T series

TSP instance	Opt	Edge-3		Edge-Ty		Edge-Tx		Edge-T	
		Ave	Std	Ave	Std	Ave	Std	Ave	Std
eil51	426	430.0	3.2	430.6	3.9	427.8	2.2	427.0	0.6
gr96	51229	54456	1190	54448	1019	51915	445	51288	121
eil101	629	673.8	10.8	676.4	10.0	641.0	5.0	630.3	2.0
lin105	14379	15643	478	15795	514	14717	236	14477	53
d198	15780	19397	506	19366	403	16300	228	15879	38
kroA200	29368	39327	1114	40006	1216	30457	313	29657	167
lin318	42029	93830	12548	80808	13552	44156	2470	43371	873
pcb442	50778	171426	11345	155100	9276	62914	2964	59071	2471

Table 2. t-test of solution quality in term of tour length

TSP instance	Edge3 – EdgeTy		Edge3 – EdgeTx		EdgeTy – EdgeTx		EdgeTx – EdgeT	
	$t(df)$	p	$t(df)$	p	$t(df)$	p	$t(df)$	p
eil51	-0.65 (55)	0.259	3.10 (51)	1.56E-03	3.43 (45)	6.61E-04	1.92 (33)	3.17E-02
gr96	0.03 (56)	0.489	10.96 (36)	2.57E-13	12.48 (39)	1.71E-15	7.45 (33)	7.36E-09
eil101	-0.97 (57)	0.169	15.09 (40)	2.06E-18	17.34 (42)	4.65E-21	10.88 (38)	1.54E-13
lin105	-1.19 (57)	0.120	9.52 (42)	2.43E-12	10.44 (40)	2.75E-13	5.43 (31)	3.09E-06
d198	0.26 (55)	0.397	30.59 (40)	1.19E-29	36.27 (45)	2.94E-35	9.98 (30)	2.42E-11
kroA200	-2.26 (57)	0.014	42.00 (33)	1.59E-30	41.65 (32)	9.44E-30	12.35 (44)	3.38E-16
lin318	3.86 (57)	1.45E-04	21.27 (31)	2.35E-20	14.57 (30)	1.88E-15	1.64 (36)	5.47E-02
pcb442	6.10 (55)	5.50E-08	50.69 (32)	1.94E-32	51.85 (34)	2.99E-34	5.45 (56)	5.75E-07

df : degree of freedom

Next, Table 3 compares the convergence speed in terms of the number of evaluations. Here the number of evaluations is defined as the number of reproduction, which is also equal to the number of crossovers. From Table 3, we can find Edge-T and Edge-Tx converges with fewer evaluations than Edge-3 and Edge-Ty; furthermore, Edge-T uses much fewer evaluations than Edge-Tx does. In most cases, Edge-T uses less than half number of evaluations that Edge-3 and Edge-Ty needs. Additionally, Edge-T gives a smaller standard deviation in the number of evaluations than other approaches. The one-tailed t-test is also given in Table 4 to test the significance of improvement in convergence speed. Here we still follow the confidence level $\alpha = 0.01$. The results of the test (Edge3 – EdgeTy) in Table 4 indicate Edge-Ty can converge faster than Edge-3 does in problem gr96, lin105, and kroA200, but this superiority does not exist in other test instances. In addition, the test (Edge3 – EdgeTx) and (EdgeTy – EdgeTx) shows that Edge-Tx is capable of a faster convergence than Edge-3 and Edge-Ty, which demonstrate the benefit of greedy manner. Finally, the test (EdgeT – EdgeTx) confirms that Edge-T can converge further faster than EdgeTx. This favorable outcome proves the utility of tabu restriction in accelerating convergence speed.

Table 3. Comparative number of evaluations

TSP instance	Edge-3		Edge-Ty		Edge-Tx		Edge-T	
	Ave	Std	Ave	Std	Ave	Std	Ave	Std
eil51	39599	3768	39730	4282	31788	4287	17766	2306
gr96	103932	8967	98328	7247	61236	5882	33811	4607
eil101	119691	18883	112961	15579	76612	10092	51660	7498
lin105	147018	13249	137509	16380	83175	11424	42045	3985
d198	653707	95779	608002	83460	284786	38178	171094	27040
kroA200	474843	57895	431967	59104	196572	19833	123077	12613
lin318	1666891	379840	1707216	411641	1473512	239780	784920	140114
pcb442	2910549	952821	3110969	979338	2423476	738469	1534216	851173

Table 4. t-test of convergence speed in terms of number of evaluations

TSP in-stance	Edge3 – EdgeTy		Edge3 – EdgeTx		EdgeTy – EdgeTx		EdgeTx – EdgeT	
	$t(df)$	p	$t(df)$	p	$t(df)$	p	$t(df)$	p
eil51	-0.13 (57)	0.450	7.50 (57)	2.37E-10	7.18 (57)	8.00E-10	15.78 (44)	4.99E-20
gr96	2.66 (55)	0.005	21.81 (50)	1.70E-27	21.77 (55)	5.27E-29	20.10 (54)	4.84E-27
eil101	1.51 (55)	0.069	11.02 (44)	1.53E-14	10.73 (49)	9.35E-15	10.87 (53)	2.10E-15
lin105	2.47 (55)	0.008	19.99 (56)	1.83E-27	14.90 (51)	1.61E-20	18.62 (35)	4.86E-20
d198	1.97 (56)	0.027	19.60 (37)	1.97E-21	19.29 (40)	3.63E-22	13.31 (52)	1.11E-18
kroA200	2.84 (57)	0.003	24.91 (35)	3.72E-24	20.68 (35)	1.66E-21	17.13 (49)	1.26E-22
lin318	-0.39 (57)	0.347	2.36 (48)	1.12E-02	2.69 (46)	5.00E-03	13.58 (46)	5.23E-18
pcb442	-0.80 (57)	0.213	2.21 (54)	1.56E-02	3.07 (53)	1.69E-03	4.32 (56)	3.20E-05

5   Conclusions and Future Work

In this work, we propose two strategies to improve the performance of edge recombination. First, inheritance of edges from alternate parents can enhance the performance of crossover. Second, a greedy manner in connecting sub-tours contributes to better results. The proposed algorithm Edge-T, based on edge recombination, is designed to verify these two strategies. The first strategy is accomplished through a penal way. Edge-T uses a tabu list to record the preceding parent and prohibit successive edge from inheriting the same parent. In addition, a greedy manner based on the second strategy is used to handle failures in Edge-T. The failure in edge recombination occurs when there is no city available to choose. This deadlock tour segment, however, can be viewed as a sub-tour. Accordingly Edge-T attempts to introduce the shortest edge to connect these sub-tours.

Several widely-used TSP instances sized from 51 cities to 442 cities are implemented to examine the performance of Edge-T. Experimental results demonstrate the capability that tabu restriction can increase the alternation of parents in inheritance. Furthermore, the performance comparisons confirm the two proposed strategies can significantly enhance Edge-T and result in better performance than Edge-3 in terms of both solution quality and convergence speed.

Much more work is needed to validate the effectiveness of the proposed strategies. First, the scale of TSP instances used in our experiments is relatively small. In addition, even though Edge-T can obtain better results than Edge-3 in these experiments, it does not achieve the optimal solution. We reason this defect due to the lack of local search. Genetic algorithms can efficiently and effectively lead the search to the promising region; however, without local search it becomes more than difficult to probe into the optimum, especially when the length of chromosome gets longer. Therefore, local search, such as 2-opt [Cro58] and Lin-Kernighan [LK73], is probably used to reinforce the search ability. Moreover, larger-scale TSP instances should be included in the benchmarks. Second, we disabled the mutation operator in our experiments for observing the behavior of crossover. This setting will cause the search to focus on exploitation, but on the other hand, exploration also needs to be considered. Hence, the operator which can enhance exploration, e.g. mutation or 2-opt, should be performed with crossover. Currently we are studying and conducting experiments on these issues. More specific analysis on the impact of the two proposed strategies upon Edge-T and applications of these two strategies to other crossovers are also underway.

References

[Cro58] G.A. Croes: A Method for Solving Traveling Salesman Problems. Operations Res. 6 (1958) 791-812

[Dav85] L. Davis: Applying Adaptive Algorithms to Domains. Proc. Int. Joint Conf. on Artificial Intelligence (1985) 162-164

[DW94] J. Dzubera and D. Whitley: Advanced Correlation Analysis of Operators for the Traveling Salesman Problems. Parallel Problem Solving from Nature-PPSN III (1994) 68-77

[GL85] D. Goldberg and R. Lingle: Alleles, Loci and the Traveling Salesman Problem. Proc. First Int. Conf. on Genetic Algorithms (1985) 154-159

[GK97] F. Glover, J.P. Kelly: Tabu Search. Kluwer Academic Publishers (1997)

[Hol75] J.H. Holland: Adaptation in Natural and Artificial Systems. University of Michigan Press (1975)

[LK73] S. Lin and B.W. Kernighan: An Effective Heuristic Algorithm for the Traveling Salesman Problem. Operations Res. 21 (1973), 498-516

[MdWS91] B. Manderick, M. de Weger, P. Spiessens: The Genetic Algorithm and the Structure of the Fitness Landscape. Proc. Fourth Int. Conf. on Genetic Algorithms (1991) 143-150

[MW92] K. Mathias and D. Whitley: Genetic Operators, the Fitness Landscape and the Traveling Salesman Problem. Parallel Problem Solving from Nature-PPSN 2 (1992) 219-228

[NK97] Y. Nagata and S. Kobayashi: Edge Assembly Crossover: A High-Power Genetic Algorithm for the Traveling Salesman Problem. Proc. Seventh Int. Conf. on Genetic Algorithms (1997) 450-457

[NK99] Y. Nagata and S. Kobayashi: An Analysis of Edge Assembly Crossover for the Traveling Salesman Problem. Proc. IEEE Int. Conf. on Systems Man and Cybernetics (1999) 628-633

[NYY00] H.D. Nguyen, I. Yoshihara, and M. Yasunaga: Modified Edge Recombination Operators of Genetic Algorithms for the Traveling Salesman Problem. Proc. IEEE Int. Conf. on Industrial Electronics, Control, and Instrumentation (2000) 2815-2820

[OSH87] I. Oliver, D. Smith, and J. Holland. A Study of Permutation Crossover Operators on the Traveling Salesman Problem. Proc. Second Int. Conf. on Genetic Algorithms (1987) 224-230

[Sta91] T. Starkweather, S. McDaniel, K. Mathias, C. Whitley, and D. Whitley: A Comparative Study of Genetic Sequencing Operators. Proc. Fourth Int. Conf. on Genetic Algorithms (1991) 69-76

[TSPLIB] http://www.iwr.uni-heidelberg.de/groups/comopt/software/TSPLIB95/

[Wat98] J. P. Watson, C. Ross, V. Eisele, J. Denton, J. Bins, C. Guerra, D. Whitley, and A. Howe: The Traveling Salesrep Problem, Edge Assembly Crossover, and 2-opt. Parallel Problem Solving from Nature-PPSN V (1998)

[WSF89] D. Whitley, T. Starkweather, and D. Fuquay. Scheduling Problems and the Traveling Salesman: the Genetic Edge Recombination Operator. Proc. Third Int. Conf. on Genetic Algorithms and Their Applications (1989)

Clustering Nominal and Numerical Data: A New Distance Concept for a Hybrid Genetic Algorithm

Laetitia Vermeulen-Jourdan, Clarisse Dhaenens, and El-Ghazali Talbi

LIFL-Université de Lille1, Bât M3-Cité Scientifique,
59655 Villeneuve d'Ascq Cedex FRANCE jourdan@lifl.fr,
http://www.lifl.fr/jourdan

Abstract. As intrinsic structures, like the number of clusters, is, for real data, a major issue of the clustering problem, we propose, in this paper, CHyGA (Clustering Hybrid Genetic Algorithm) an hybrid genetic algorithm for clustering. CHyGA treats the clustering problem as an optimization problem and searches for an optimal number of clusters characterized by an optimal distribution of instances into the clusters. CHyGA introduces a new representation of solutions and uses dedicated operators, such as one iteration of K-means as a mutation operator. In order to deal with nominal data, we propose a new definition of the cluster center concept and demonstrate its properties. Experimental results on classical benchmarks are given.

1 Introduction

Clustering is used to identify classes of objects sharing common characteristics and its methods can be applied to many human activities and particularly to the automatic decision making problem. The data clustering or unsupervised classification, can isolate similarities and differences in a database and make groups of similar data which are called classes, groups or clusters. It can reveal some intrinsic structures (e.g. the number of clusters). There exists many clustering methods like graph-based ones, model-based ones, genetic algorithm-oriented ones, distance based approaches or their hybridizations. Most of these methods require as an input the number of clusters. This requirement is a major problem for real-life problems, where the number of clusters is not known in advance.

Genetic algorithms have been successfully applied to partitioning problems [10] and in particular to clustering problems. In [12], the authors couple the fuzzy K-means algorithm with a GA, where one iteration of the Fuzzy K-means is used to compute the fitness of the classification. However, all those algorithms require as input the number of clusters.

In this paper, we present an hybrid genetic algorithm using a specific encoding with dedicated operators. The hybridization consists in using a step of the K-means algorithm (a single iteration) as one of these operators. We recall that K-means is devoted to deal with numerical data, and conversely we aim at dealing with nominal data. Therefore, in order to realize the proposed hybridization,

J. Gottlieb and G.R. Raidl (Eds.): EvoCOP 2004, LNCS 3004, pp. 220–229, 2004.

we need to redefine an iteration of the K-means algorithm. In addition, we propose a new definition for the cluster center concept. An associated distance is also proposed.

Section 2 presents the K-means algorithm, the definition of the center concept for nominal data we introduce and the proposed associated distance. Section 3 presents CHyGA, its encoding and the dedicated operators. Section 4 provides experimental results for several classical numerical and nominal datasets.

2   Clustering and Center Concept

Clustering aims to group similar objects into clusters which can be described by theirs centers. Each object is described by a set of attributes. Each attribute A_i has a definition domain Ω and takes a value in this domain.

The K-means algorithm is one of the most famous algorithm for clustering [5]. We firstly present the classical algorithm dedicated to numerical data, then we introduce a new definition of the center concept more adapted to nominal data. We also introduce the associated distance.

2.1   The K-Means Algorithm

The K-means algorithm is an iterative procedure where an iteration as described below, consists in improving an existing partition:

> **Input:** Partition P of k clusters: $C_1, ..., C_k$;
> Compute $Center(C_1), ..., Center(C_k)$;
> Remove all objects from all cluster;
> **for** each object O_i **do**
> > Let C_j, $(j \in [1, k])$ be the cluster whose center is the closest to O_i;
> > Assign O_i to C_j;
>
> **end for**
> Compute the resulting new partition $P = C_1, ..., C_l(l \leq k)$;
> Remove all empty clusters.

Major drawbacks of K-means algorithm is that it often terminates on a local optimum and it works only on numerical values because it minimizes a cost function calculating means of clusters. Moreover, it needs to compute centers. The center of a cluster is easy to define on numerical values because the mean makes sense, but for nominal data it is not so simple.

2.2   A New Definition for the Cluster Center Concept

In some works, authors have proposed some definitions for the center of categorical or nominal data. For example, Huang proposes to compute the center of a cluster by using the mode of a set [13].

Definition 1. Let X be a set of n nominal objects $X_1, .., X_n$ described by attributes $A_1, ..., A_m$, Ω the set of all possible combinations of values of the attributes.

A mode of X is a vector $Q \in \Omega$, $Q = [q_1 q_2...q_m]$ that minimizes $D(Q, X) = \sum_{j=1}^{n} d(X_j, Q)$, where $d(A, B)$ is a simple matching between 2 objexts, as defined in [18]. Q is not necessarily an element of X.

This definition has two drawbacks: first, Q is not always unique; second, if we consider an attribute having the following values $Y, ?, Y, N, Y, N, ?$ the mode will choose the value Y which has a frequency of 3, but is it really significant ?
We decide not to use such a center election. We propose here to consider a center election based on a majority vote (frequency $> 1/2$). When there is no satisfiable value, none is chosen and we use a partially defined center. We will use the notation * to denote values of attributes for which there is no satisfiable candidate.

Definition 2. Let X be a set of n nominal objects described by attributes $A_1, ..., A_m$.
The center of X is the vector $Q = [q_1 q_2...q_m]$ with $q_i \in \Omega \cup \{*\}$ which minimizes $D(Q, X) = \sum_{j=1}^{n} d_v(X_j, Q)$ where a possible distance measure for d_v is defined in paragraph 2.3. Q is not necessarily an element of X and is unique.

The defined center will be called a partial center and represents an hyperplane whereas for commonly used definition, a center is reduced to a single point of the space $\mathbb{R}^m$ (like for mode). The dimension of the hyperplane in $\mathbb{R}^m$ is the total number of attributes - number of determined attributes of the center.

2.3   A Proposed Associated Distance Measure

In order to realize a clustering with K-means, we have to define a distance between an object and a partially defined center. The proposed distance is based on the Hamming distance, but had to be adapted in order to deal with the partial center concept.
Let $d_v(O, C)$ be the distance between an object O and a center C. We define d_v as the sum over every attributes ($d_v = \sum_{i=1}^{m} d_\sigma(O_i, C_i)$) as follow:

$$\forall x \text{ and } y, \quad d_\sigma(x, y) = \begin{cases} 0 \text{ if x=y} \\ 1/2 \text{ if x or y equals to * (x$\neq$y)} \\ 1 \text{ if x $\neq$ y} \end{cases} \qquad (1)$$

Let us demonstrate the properties of d_v and show that it is a distance.

Proof. Positivity
By construction, d_σ is always positive or equal to zero then
$d_v(O, C) = \sum_i d_\sigma(O_i, C_i) \geq 0$.

Proof. Symmetry
$d_v(O, C) = \sum_i d_\sigma(O_i, C_i)$ then if d_σ is symmetrical, d_v is symmetrical.
Let us consider the different cases:
If $d_\sigma(O_i, C_i) = 0 \Longrightarrow O_i = C_i \Longleftrightarrow C_i = O_i$ then $d_\sigma(C_i, O_i) = 0$.
If $d_\sigma(O_i, C_i) = \frac{1}{2} \Longrightarrow C_i = *$ (O_i always defined) then $d_\sigma(C_i, O_i) = \frac{1}{2}$.
If $d_\sigma(O_i, C_i) = 1 \Longrightarrow O_i \neq C_i$ then $C_i \neq O_i$ then $d_\sigma(C_i, O_i) = 1$.
d_σ is symmetrical then d_v is symmetrical.

Proof. Triangular inequality

Let O_1, O_2, O_3 be three objects that can be centers.

We want to show that $d_v(O_1, O_2) \leq d_v(O_1, O_3) + d_v(O_3, O_2)$.

We must show that $\sum_i d_\sigma(O_{1_i}, O_{2_i}) \leq \sum_i d_\sigma(O_{1_i}, O_{3_i}) + \sum_i d_\sigma(O_{3_i}, O_{2_i})$.

Let $A = \sum_i d_\sigma(O_{1_i}, O_{2_i})$ et $A_i = d_\sigma(O_{1_i}, O_{2_i})$.

Let $B = \sum_i d_\sigma(O_{1_i}, O_{3_i}) + \sum_i d_\sigma(O_{3_i}, O_{2_i})$ and $B_i = d_\sigma(O_{1_i}, O_{3_i}) + d_\sigma(O_{3_i}, O_{2_i})$.

We compute attribute by attribute. For attribute i, three cases may appear:

1. If $O_{1_i} = O_{2_i}$ then $A_i = 0$:
   - If $O_{1_i} = O_{3_i}$ then $O_{2_i} = O_{3_i}$ then $B_i = 0$
   - If $O_{1_i} \neq O_{3_i}$ then $B_i + = \frac{1}{2}$ or 1 and then $O_{3_i} \neq O_{2_i}$ then $B_i + = \frac{1}{2}$ or 1
   $\implies B_i = 0, 1$ or 2. In this case $A_i \leq B_i$.

2. If $O_{1_i} \neq O_{2_i}$ and $\neq *$ then $A_i = 1$:
   - If $O_{1_i} = O_{3_i}$ then $B_i + = 0$ but $O_{3_i} \neq O_{2_i}$ then $B_i + = 1$.
   - If $O_{1_i} \neq O_{3_i}$ then if $O_{3_i} = *$ then $B_i = \frac{1}{2} + \frac{1}{2}$ else $B_i = 1 + (0 \ or \ 1)$
   $\implies B_i = 1$ or 2. In this case, $A_i \leq B_i$.

3. If $O_{1_i} \neq O_{2_i}$ and one of them equals to $*$. As the distance is symmetrical, let $O_{1_i} = *$ then $A_i = \frac{1}{2}$:
   - If $O_{1_i} = O_{3_i}$ then $B_i + = 0$ but $O_{3_i} \neq O_{2_i}$ then $B_i + = \frac{1}{2}$.
   - If $O_{1_i} \neq O_{3_i}$ then $B_i = \frac{1}{2} + (0 \ or \ 1)$.
   - If $O_{3_i} = *$ then $B_i = \frac{1}{2}$.
   $\implies B_i \geq \frac{1}{2}$. In this case $A_i \leq B_i$.

To conclude, for all objects O_1, O_2, O_3:

$\forall i \ A_i \leq B_i \implies \sum_i A_i \leq \sum_i B_i \implies A \leq B$

$\sum_i d_\sigma(O_{1_i}, O_{2_i}) \leq \sum_i d_\sigma(O_{1_i}, O_{3_i}) + \sum_i d_\sigma(O_{3_i}, O_{2_i})$

$d_v(O_1, O_2) \leq d_v(O_1, O_3) + d_v(O_3, O_2)$

Thus d_v is positive, symmetrical and transitive so, we can conclude that d_v is a distance.

2.4 Properties of the Center

In this section, we must verify that the center Q defined in Section 2.2 respects the fundamental properties of a center regarding the distance described in Section 2.3.

Theorem 1. The function $D(X, Q) = \sum_j d_v(X_j, Q)$ is minimized by the center Q defined in definition 2.

Proof. Let n be the number of objects in the cluster X, let m be the number of attributes of an object. Let d_v be the previous defined distance ($d_v = \sum_i d_\sigma(O_i, C_i)$). Let Q be a center of the cluster: $\sum_{j=1}^n d_v(X_j, Q) = \sum_{j=1}^n \sum_{i=1}^m d_\sigma(X_{j_i}, q_i)$.

As $\forall i \in [1, m] \ d_\sigma(X_{j_i}, q_i) \geq 0$, if each element of the sum is minimal $D(X, Q)$ is minimal.

Let n_{q_i} be the frequency of the value q_i of the attribute i chosen to be the representative center. Recall that the distance between $*$ and an attribute is always 1/2. As n_{q_i} is the frequency of $q_i : n - n_{q_i} \geq 0$.

For each $d_\sigma(X_{i_j}, q_j)$ two cases are possible :

1. Let the ith attribute of the center be determined. By definition of the center, the attribute has a frequency greater than the half of the voices ($n_{q_i} > \frac{n}{2}$). As $d_\sigma(X_{j_i}, q_i) = n - n_{q_i}$ then $n - n_{q_i} \leq \frac{n}{2}$. Thus $d_\sigma(X_{j_i}, q_i)$ obtained by the majority vote is minimal.

2. Let the ith attribute of the center be undetermined. There exists no value representing the majority for this attribute. So $\forall q_i, n - n_{q_i} > n/2$ then $d(X_{j_i}, q_i)$ obtained with the * is minimal.

Hence, the proposed definition of center minimizes the presented intra-cluster distance.

3 CHyGA

The clustering problem is NP-hard [19], and may be treated as an optimization problem. The literature shows that GAs are well adapted to explore the very large search space of this problem [14,17]. Figure 1, shows the different stages of CHyGA, the genetic algorithm we propose. We present here the main characteristics of this algorithm.

3.1 Encoding

For clustering or grouping problems, there exist different representations [3,10, 16,20]. For example the group number representation which indicates for each object the group it belongs to [16]. All those representations have advantages and drawbacks.

In CHyGA, we choose to use a double hybrid representation (see Figure 2) which merges the group number representation and a description of clusters. This representation allows to find very quickly information on object's affectations and cluster's composition.

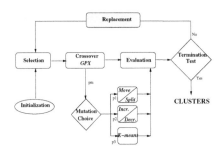

Fig. 1. Stages of CHyGA.

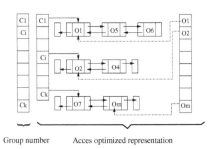

Fig. 2. The encoding.

3.2   Fitness Function

In the literature, several cluster validity indicators have been used [2,8,9]. To measure the quality of the proposed clustering, we use as fitness function the criterion of Calinski and Harabasz (CH) that measures the good distribution of the objects [6].

Given k clusters, n features, let χ_j be the j^{st} cluster with j=1...k.

Let m be the mean vector $m = (m_1\ m_2\ \cdots\ m_n)$ and m_j the vector of means for the j^{st} cluster $m_j = (m_{j_1}\ m_{j_2}\ \cdots\ m_{j_n})$. X_i is an element of the cluster χ_j and $(X_i - m_j)^t$ is the transposed vector. Then the Calinski and Harabasz criterion is:

$$CH = \frac{(n-k) \times trace(\sum_{j=1}^{k} n_j (m_j - m)(m_j - m)^t)}{(k-1) \times trace(\sum_{j=1}^{k} \sum_{X_{i \in \chi_j}} (X_i - m_j)(X_i - m_j)^t)} \quad (2)$$

3.3   Operators

Crossover

In CHyGA, we use the GPX crossover (Greedy Partition Crossover) [11] which is crossover dedicated to partition problems. Our representation is well adapted to this operator, as elements of a same cluster are grouped.

Mutations

Our GA uses 5 different mutation operators [1]. For clustering problems mutation operators must able to introduce new groups and to remove existing groups. In [10], Falkenauer shows that the mutation of a single affectation (object/group) is not enough because it doesn't increase significantly the fitness and the new candidate solution will be quickly lost from the population. Our GA uses two standard mutations for grouping problems : split and move [7], two specifically designed mutation operators adapted to clustering and a single iteration of K-means.

The split operator selects some objects in a particular cluster and moves these objects to a new cluster. We also define two other operators that are more specific to the clustering. The first operator, called "increase", is able to increase the number of clusters. This mutation operator chooses the cluster with the maximum internal variance and splits it into two clusters. Objects are reallocated between the two new clusters. Our second specifically designed mutation operator, "decrease", aims at merging two clusters into one. This operator detects the two closest centers and groups the corresponding clusters into one.

We also use one iteration of the K-means [14] algorithm as a mutation operator to make a local search and improve the quality of a solution. It takes as input the distribution given by a solution, calculates centers of clusters and reassigns objects to clusters whose center are the closest.

At the end of the genetic search, we try to improve the quality of all the solutions found with a local search. The local search consists in applying an iteration of the K-means algorithm.

4   Experimental Results

We have experimented CHyGA on some common datasets available at the UCI repository [4], or in the articles of Bandyopadhyay [2] and Ruspini [21]. Specificities of each dataset used to evaluate the performances of the proposed algorithm are summed up in Table 1. It indicates the name of the dataset its number of instances (n), the number of attributes (m), the original number of classes (k), the percentage of missing data (l). Therefore a dataset is summed up by **Name** $(\mathbf{n, m, k, l})$. Table 1 also indicates the CH value obtained with the initial repartition of the instances in their classes as all the datasets are basically intended to the classification data mining task. The steady state ChyGA parameters are: population size: 200, mutation rate: 0.9, crossover rate: 0.5 and it stops when the best individual has not changed since 100 generations.

Table 1. Descriptive statistics of the results obtained by CHyGA on different numerical and nominal datasets.

Numerical Datasets			Numerical Datasets			Nominal Datasets		
AD_5_2			**Iris**			**Lung**		
(250, 2, 5, 0)			**(150, 4, 3, 0)**			**(32, 57, 2, 4.13)**		
CH = 387.75			**CH = 486.32**			**CH =9.609**		
Criteria	# cluster		Criteria	# cluster		Criteria	# cluster	
Mean	386.20	5	Mean	526.42	3	Mean	23.38	2
Max.	387.75	5	Max.	559.25 (3)	4	Max.	23.68	2
σ	0.18	0	σ	11.78	0.82	σ	0.14	0
Med.	386.44	5	Med.	527.83	3	Med.	23.34	2
AD_4_3			**Cancer**			**Vote**		
(400, 3, 4, 0)			**(683, 9, 2, 0)**			**(435, 16, 2, 0.22)**		
CH = 3207.41			**CH = 912.20**			**CH =455.865**		
Criteria	# cluster		Criteria	# cluster		Criteria	# cluster	
Mean	3207.41	4	Mean	1025.81	2	Mean	508.79	2
Max.	3207.41	4	Max.	1025.81	2	Max.	522.34	2
σ	0	0	σ	0	0	σ	9.76	0
Med.	3207.41	4	Med.	1025.81	2	Med.	507.52	2
Ruspini			**Diabetes**			**Breast Cancer**		
(75, 2, 4, 0)			**(768, 8, 3, 0)**			**(286,9, 2, 0.34)**		
CH = 425.32			**CH = 24.29**			**CH =212.251**		
Criteria	# cluster		Criteria	# cluster		Criteria	# cluster	
Mean	425.32	4	Mean	1136.18	3	Mean	243.63	2
Max.	425.32	4	Max.	1142.49	3	Max.	247.49	2
σ	0	0	σ	6.98	0	σ	2.99	0
Med.	425.32	4	Med.	1139.2	3	Med.	243.91	2

Table 1 also indicates for each dataset some descriptive statistics of the results obtained by CHyGA. We indicate for the Calinski and Harabasz criteria (CH) the mean of the best results obtained over ten runs, the maximum, the standard deviation (σ) and the median. Information is given about the number of discovered clusters.

We can observe that CHyGA is able to discover a number of clusters equals to the original number of classes. Only on one dataset, Iris, it sometimes obtains, for the best solution of an execution, a number of clusters of four instead of three. However, for these data, the best solution ever obtained for CH has been found for a number of clusters equal to three (559.258(3)).

Concerning the robustness of the method, we can observe that the method is very robust on some of the numerical datasets where the standard deviation of the best results obtained by CHyGA over the runs is equal to zero.

In addition, we compare our method with some classical methods: K-means,

Table 2. Best results obtained with different method on classical dataset from UCI.

Data	Criteria	CHyGA	K-means	Kmed.	Single Link	Compl. Link
AD_5_2	Intra	**5.06**	5.17	6.08	93.90	5.62
	Inter	218.85	218.80	218.02	**218.89**	218.88
	Calinski	**388.12**	386.58	163.92	11.74	6.25
	σ	0	28.84	24.68	-	-
AD_4_3	Intra	**6.54**	6.63	8.75	6.55	19.67
	Inter	410.88	410.88	411.10	**410.83**	410.99
	Calinski	**3207.41**	3206.68	1720.86	3190.08	0.57
	σ	0	385.02	313.31	-	-
Ruspini	Intra	**38.75**	**38.75**	54.18	40.57	162.15
	Inter	19782.42	19782.42	**19787.28**	19783.07	19782.52
	Calinski	**425.32**	**425.32**	263.92	422.40	2.63
	σ	0	161.56	74.2	-	-
Iris	Intra	**2.36**	**2.36**	2.86	2.69	2.50
	Inter	46.6264	46.6264	**46.93**	46.62	46.62
	Calinski	**559.26**	**559.26**	322.92	343.85	42.35
	σ	11.78	1.98	108.55	-	-
Cancer	Intra	14.14	17.08	13.42	28.16	20.03
	Inter	235.31	235.31	237.68	235.17	235.34
	Calinski	**1026.06**	1008.74	1201.66	298.09	12.06
	σ	0	0	385.02	-	-
Diabetes	Intra	**246.56**	247.44	276.17	454.01	264.00
	Inter	105763.16	105763.25	105736.98	**105763.90**	105760.35
	Calinski	**1135.07**	133.97	276.17	20.19	0.95
	σ	6.98	17.56	47.51	-	-

Kmediods, Single link [15] and Complete link [15]. To make a fair comparison, we give as input, for all the classical methods, the number of clusters found by CHyGA. For non deterministic methods (K-means, Kmediods), we ran ten times the algorithms. Table 2 indicates for each method, the best solution found, the standard deviation (σ) of the solution given by the algorithm for CH. As the other algorithms are not designed to optimize CH, we also compare the regarding objectives often used in clustering: minimizing the intra cluster distance and maximizing the inter cluster distance. Table 2 also indicates their values.

For numerical datasets, we can observe that our method finds for all the datasets the best solution of CH also found sometimes by the K-means algorithm. For intra cluster distance which measures the compactness of the cluster, we can observe that CHyGA finds for each dataset the smaller value which is for two datasets, Ruspini and Iris, also obtained by K-means. For inter cluster criteria which measures the separation of the obtained clusters, we remark that even if CHyGA doesn't find the best value of the criteria, the value obtained is really closed to the best found. Indeed, when we look at the error made by CHyGA in comparison with the best found, this error is less than 0.02% for Ruspini dataset, 0.66% for Iris dataset and 0.0006% for Diabetes dataset.

For the three stochastic methods, we indicate the standard deviation of the solutions obtained over the ten executions of the algorithms (σ). We can observe that CHyGA has the smaller standard deviation except for the Iris dataset.

Concerning the computational time, it is obvious that our method is longer than a simple K-means algorithm because we use a single iteration of it as a local search in our hybridization. Methods such as Single or Complete link are faster than our method on small datasets but longer on larger datasets (Diabetes for example).

Hence, those results show that CHyGA has best or at least comparable performances than the best clustering methods. Nevertheless, it is important to recall that CHyGA does not require as input the number of clusters to look for and it can deal with nominal or numerical data.

5   Conclusion

This paper has proposed a specific hybrid Genetic Algorithm for the clustering problem: CHyGA. This algorithm is very interesting because it makes clustering with no indication on the number of clusters by using the Calinski and Harabasz criteria. We have used a specific encoding and an one iteration K-means to hybridize the algorithm. To deal with nominal data we proposed a new center conception and an associated distance. Results obtained are very encouraging. We obtained on both numerical and nominal data the same number of clusters than the number of given classes.

This algorithm works for an unknown number of clusters and the final population is compound of different solutions with different number of clusters. Here, only the best solution has been used for the evaluation, but it could be interesting to look at several good solutions obtained thanks to CHyGA.

A multi-criteria approach would be interesting to look for the best compromise between different quality criteria that may be used.

References

1. T. Back, D.B. Fogel, and Z. Michalewicz, editors. *Handbook of Evolutionary Computation*. Oxford University Press, 1997.
2. S. Bandyopadhyay and U. Maulik. Genetic clustering for automatic evolution of clusters and application to image classification. *Pattern Recognition*, 35:1197–1208, 2002.
3. J.C. Bezdeck, S. Boggavaparu, L.O. Hall, and A. Bensaid. Genetic algorithm guided clustering. In *Proc. of the First IEEE Conference on Evolutionary Computation*, pages 34–38, 1994.
4. C.L. Blake and C.J. Merz. Uci repository of machine learning databases. http://www.ics.uci.edu/~mlearn/MLRepository.html, 1998. University of California, Irvine, Dept. of Information and Computer Sciences.
5. L. Bottou and Y. Bengio. Convergence properties of the K-means algorithms. In G. Tesauro, D. Touretzky, and T. Leen, editors, *Advances in Neural Information Processing Systems*, volume 7, pages 585–592. The MIT Press, 1995.
6. T. Calinski and J. Harabasz. A dendrite method for cluster analysis. *Communications in statistics*, 3(1):1–27, 1974.
7. R.M. Cole. Clustering with genetic algorithms. Master's thesis, University of Western Australia, Australia, 1998. http://citeseer.nj.nec.com/cole98clustering.html.
8. D.L. Davies and D.W. Bouldin. A cluster separation measure. *IEEE Transactions on Pattern Analysis and Machine Intelligence*, 1, 1979.
9. J.C. Dunn. A fuzzy relative of the isodata process and its use in detecting compact, well-seperated clusters. *Journal of Cybernetics*, 3(3):32–57, 1973.
10. E. Falkenauer. *Genetic Algorithms and Grouping Problems*. John Wiley, 1998.
11. P. Galinier and JK. Hao. Hybrid evolutionary algorithms for graph coloring. *Journal of Combinatorial Optimization*, 3:379–397, 1999.
12. L. O. Hall, I. B. Oezyurt, and J. C. Bezdek. Clustering with a genetically optimized approach. *IEEE Transactions on EC*, 3(2):103–112, July 1999.
13. Z. Huang. Extensions to the k-means algorithm for clustering large data sets with categorical values. *Data Mining and Knowledge Discovery*, 2(3):283–304, 1998.
14. A.K. Jain, M.N. Murty, and P.J. Flynn. Data clustering: A review. *ACM Computing Surveys*, 31(3):264–323, September 1999.
15. N. Jardine and C.J. van Rijsbergen. The use of hierarchical clustering in information retrieval. *Information Storage and Retrieval*, 7(5):217–240, 1971.
16. D.R. Jones and M.A. Beltramo. Solving partitioning problems with genetic algorithms. In *Proc. of the Fourth International Conference on Genetic Algorithms*, pages 442–449. Morgan Kaufman Publishers, 1991.
17. L. Jourdan, C. Dhaenens, E.G. Talbi, and S. Gallina. A data mining approach to discover genetic and environmental factors involved in multifactoral diseases. *Knowledge Based Systems*, 15(4):235–242, May 2002.
18. L. Kaufman and P. Rousseuw. *Finding Groups in Data- An Introduction to Cluster Analysis*. Wiley Series in Probability and Mathematical Sciences, 1990.
19. G. L. Liu. *Introduction to combinatorial Mathematics*. McGraw Hill, 1968.
20. Z. Michalewizc. *Genetic Algorithms + Data Structures = Evolution Programs*. Springer-Verlag, third, revised and extend edition, 1996.
21. E.H. Ruspini. Numerical methods for fuzzy clustering. *Inform. Sci.*, 2:319–350, 1970.

On Search Space Symmetry in Partitioning Problems

Benjamin Weinberg and El-Ghazali Talbi

Université des Sciences et Technologies de Lille LIFL - UPRESA 8022 CNRS
Bâtiment M3 Cité Scientifique, 59655 Villeneuve d'Ascq CEDEX, France
{weinberg,talbi}@lifl.fr

Abstract. Many problems consist in splitting a set of objects so that each part verifies some properties. In practice, a partitioning is often encoded by an array mapping each object to its group numbering. In fact, the group number of a object does not really matter, and one can simply rename each part to obtain a new encoding. That is what we call the symmetry of the search space in a partitioning problem. This property may be prejudicial for methods such as evolutionary algorithms (EA) which require some diversity during their executions.

This article aims at providing a theoretical framework for breaking this symmetry. We define an equivalence relation on the encoding space. This leads us to define a non-trivial search space which eliminates symmetry. We define polynomially computable tools such as equality test, a neighborhood operator and a metric applied on the set of partitioning.

1 Introduction

In many fields, a broad range of problems, graph coloring problem and bin packing problem in operational research, the clustering problem in knowledge discovery domain (etc), can be formalized as partitioning problems [2]. Formally, all those problems consist in partitioning a set U of objects into mutually disjunctive subsets U_i (i.e., such that: $\bigcup U_i = U$ and $U_i \bigcap U_j = \emptyset, i \neq j$). These problems use the same search space but may differ on specific constraints and/or objective functions.

In partitioning problems, the representation of a solution is an important issue and more difficult than it seems to be. Generally, a solution is provided by a mapping called the group-numbering: $S : U \rightarrow \{1, 2, \ldots N\}$, where N is the size of U. This mapping provides the number of its group for each element of U. Technically, such a mapping is easy to encode by an array. But this representation leads to difficulties for some algorithms. Indeed, two different mappings may encode the same solution (see Figure 1). That is why some classic genetic algorithms with classic operators, like the one point, the two points or the uniform cross-over, are non-performing on such problems [2]. Indeed, with such cross-over operators, two "equal" configurations may provide different configurations, so the good properties may be non-conserved in offspring during the algorithm.

J. Gottlieb and G.R. Raidl (Eds.): EvoCOP 2004, LNCS 3004, pp. 230–240, 2004.

A	B	C	D	E	F	G	H
1	2	3	4	2	2	4	1

A	B	C	D	E	F	G	H
2	3	4	1	3	3	1	4

1	A H
2	B E F
3	C
4	D G

1	D G
2	A H
3	B E F
4	C

This figure presents two mappings which represent the same partition-ing of the set {A, B, ... H}. In the left of the figure, the mappings are classically shown as arrays. In the right side of the figure, the mappings are rewritten by gathering elements by set. This double representation is the main idea of the encoding presented in 2.2.

Fig. 1. Mapping encoding problem.

Formally, breaking the symmetry is not necessary for some applications by designing dedicated operators like Greedy Procedure cross-over (GPX) [3] or permutation cross-over [4] on the graph coloring problem. However, it will become essential when diversification mechanisms (like sharing and crowding [5]) are used or when fitness landscape (like correlation length and roughness[7]) is studied. In both cases, we need a metric to estimate the difficulty of a problem.

In the first case, the fitness of a configuration of a population is more or less penalized with regard to its distances with the other configuration of the population. In the second case, we observe the variation of the fitness of a configuration the distance to a reference configuration (classically the optimum). The purpose of this article is to provide some theoretical and practical tools, therefore we state a formalization of the search space.

In section 2, we present the definition and nomenclature we use; we also introduce an equivalence relation. In section 3, we build the search space as the quotient of a set by the previously defined equivalence relation. In section 4, we define a polynomially computable metric operating on such a search space. In section 5, we make some concluding remarks and suggest our future works.

2   Formalization

In this section, we first state the definitions and notations. We also suggest an encoding for partitioning which simplifies some algorithms. Then, we present an equivalence relation which is the keystone of this paper.

2.1   Definitions and Nomenclature

Definition 1. *Given U a set of objects, a* group-numbering *is any mapping from U to $\{1, 2, \ldots N\}$.*

The set of all group-numberings is denoted $\mathcal{R}$.

Definition 2. *For any given group-numbering $\mathcal{A}$ and any object x in U, we call the* group *of x the image of x by $\mathcal{A}$ (i.e. $\mathcal{A}(x)$).*

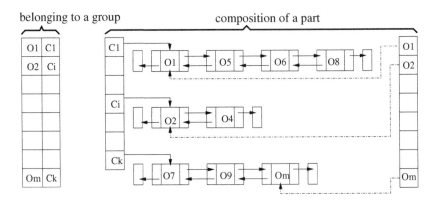

Technically the number k of non-empty parts is stored and only the first k labels are used.

Fig. 2. Example of encoding.

Definition 3. *For any given group-numbering $\mathcal{A}$ and any integer i in $\{1, 2, \ldots N\}$, we call i^{th} part of $\mathcal{A}$ the set of preimages of i by $\mathcal{A}$ (i.e. $\mathcal{A}^{-1}(i)$).*

We denote $\mathcal{A}_\sim$ a partitioning defined by a group-numbering $\mathcal{A}$. With this definition, we allow empty parts in a partitioning.

Definition 4. *A partitioning $\mathcal{A}_\sim$ is said a* refining *of a partitioning $\mathcal{B}_\sim$, if all parts of the former are included in a part of the latter.*

$$\forall i \in \{1 \ldots N\} \ \exists j \in \{1, 2, \ldots N\} \ \mathcal{A}^{-1}(i) \subset \mathcal{B}^{-1}(j)$$

Reciprocally, $\mathcal{B}_\sim$ is said a disrefining *of $\mathcal{A}_\sim$.*

2.2   Encoding

Since it is interesting to examine objects gathered by parts, we propose an encoding using double-chained lists. Each list contains the objects of a part. We also use two arrays indexed by the objects of the problem. The first one is the direct encoding of the group-numbering. It provides for each object the number of its group. The second array provides for each object an access to its corresponding link in the chained lists (see Figure 2). With this encoding the complexity of changing the part of one object is still in $O(1)$.

2.3   Equivalence Relation

Two group-numberings encode the same information if they represent the same partitioning. Intuitively, the first group-numbering is a relabeling of parts of the second one. Formally, we can observe a bijection between the parts of the two group-numberings.

$$\exists\, \sigma : \{1, 2, \ldots N\} \rightarrow \{1, 2, \ldots N\}, \forall x \in U, \sigma(\mathcal{A}(x)) = \mathcal{B}(x)$$

Definition 5. *We define the relation* $\sim$ *as follows:*
$\mathcal{A} \sim \mathcal{B}$ *if there exists a bijection* $\sigma : \{1, 2, \ldots N\} \rightarrow \{1, 2, \ldots N\}$*, so that all objects* x *of* U *verify* $\sigma(\mathcal{A}(x)) = \mathcal{B}(x)$*.*

Let us verify that $\sim$ is an equivalence relation

- Reflexivity: Let $\mathcal{A}$ be a group-numbering. We have obviously $id(\mathcal{A}(x)) = \mathcal{A}(x)$, where id is the identity. As id is obviously a bijection, we deduce that $\mathcal{A} \sim \mathcal{A}$.
- Symmetry: Let $\mathcal{A}$ and $\mathcal{B}$ be two group-numberings. Assume that $\mathcal{A} \sim \mathcal{B}$. So, there exists σ so that for all $x \in U$, we have $\sigma(\mathcal{A}(x)) = \mathcal{B}(x)$. σ is a bijection thus σ^{-1} exists and is also a bijection. We apply σ^{-1} to the definition of $\mathcal{A} \sim \mathcal{B}$, This leads to $\sigma^{-1}(\sigma(\mathcal{A}(x))) = \sigma^{-1}(\mathcal{B}(x))$. So, $\sigma^{-1}(\mathcal{B}(x)) = \mathcal{A}(x)$ thus $\mathcal{B} \sim \mathcal{A}$.
- Transitivity: Let $\mathcal{A}$, $\mathcal{B}$ and $\mathcal{C}$ be three group-numberings. Assume $\mathcal{A} \sim \mathcal{B}$ and $\mathcal{B} \sim \mathcal{C}$. We know by definition, there exists σ_1 (resp. σ_2) so that for all $x \in U$, we have $\sigma_1(\mathcal{A}(x)) = \mathcal{B}(x)$ (resp. $\sigma_2(\mathcal{B}(x)) = \mathcal{C}(x)$). Hence for all $x \in U$, we have $\sigma_2 \circ \sigma_1(\mathcal{A}(x)) = \sigma_2(\mathcal{B}(x)) = \mathcal{C}(x)$. As σ_1 and σ_2 are bijections, $\sigma_2 \circ \sigma_1$ is a bijection too. This leads to $\mathcal{A} \sim \mathcal{C}$.

3   The $\sim$ Quotient Space

In this section, we present the search space as the quotient of the set of group-numberings by equivalence relation $\sim$. Then, we present the test of equality in the space and the elementary neighborhood operator.

3.1   Building a Quotient Space

We build the search space $\mathcal{O}$ as the quotient of group-numbering by the relation $\sim$. Given a group-numbering $\mathcal{A}$, the partitioning $\mathcal{A}_\sim$ is the set of group-numberings equivalent to $\mathcal{A}$ (also called the equivalence class of $\mathcal{A}$). Therefore, we can write:

$$\mathcal{O} = \{partitionings\} = \mathcal{R}_{/\sim} = \{\mathcal{A}_\sim | \mathcal{A} \in \mathcal{R}\}$$

Practically, there is $k!$ equivalent group-numberings for a single partitioning with k parts. Due to the size of such a class, one cannot directly manipulate it. Therefore, it is necessary to find an efficient algorithm to determine if a group-numbering represents a given partitioning.

The first idea is to define a canonical group-numbering for any partitioning. For instance, this can be done by sorting elements in each part, then sorting each part by its first element.

Marino *et al.* use another strategy in [6]. They define a set of reference objects, which are always in different parts. This technique is applied to the graph coloring problem. In this case, the reference set is a clique as large as possible[1]. However, the reference set is generally hard to define on partitioning problems.

Due to the unique representation of a partitioning these methods are enough to test the equality. But it is not possible to extend them to more complex tools like distance.

3.2   Operating on the Quotient Space

In this section, we present some basic operations, i.e. the equality test, and the neighborhood operator, which moves an object from a part to another.

Equality Test. To test the equality of two partitionings provided by two different group-numberings, we have to verify the equivalence of the group-numberings. Therefore, we use an algorithm in $O(N)$ in worst case (see Algorithm 1). This algorithm exploits the encoding defined in subsection 2.2. In Algorithm 1, the first test is needed to distinguish equal partitionings from the case where $\mathcal{A}_\sim$ is a refining of $\mathcal{B}_\sim$.

Algorithm 1 Equality test of two partitionings given by representative group-numberings $\mathcal{A}$ and $\mathcal{B}$.

---

  require *sigmaOfI*: INTEGER // unique candidate for $\sigma(i)$
  if $\mathcal{A}$ and $\mathcal{B}$ have not the same number of non-empty parts **then**
    return FALSE
  end if
  for $i \in \{1 \ldots$ number of non-empty parts of $\mathcal{A}\}$ **do**
    let x be $\in \mathcal{A}^{-1}(i)$
    sigmaOfI $= \mathcal{B}(x)$
    for $t \in \mathcal{A}^{-1}(i)\backslash\{x\}$ **do**
      if $\mathcal{B}(t) \neq$ *sigmaOfI* **then**
        return FALSE
      end if
    done
  done
  return TRUE

Elementary Neighborhood Operator. This operator, namely *move_object_to_part* (MOTP), consists in moving an object from its part to another. This operator needs three parameters: the partitioning $\mathcal{A}_\sim$, the object

---

[1] Maximum clique problem is also an NP-hard problem.

$$COG : \quad \mathcal{R}, U, \quad \mathbb{N} \quad \longrightarrow \mathcal{R}$$

$$\updownarrow \qquad\qquad \updownarrow$$

$$MOPT : \mathcal{O}, U, P(V) \longrightarrow \mathcal{O}$$

The hook vertical arrow designates the canonical surjection (i.e. a partitioning $\mathcal{A}$ is mapped to its equivalence class $\mathcal{A}_\square$). Operating on partitionings can be viewed as operating on any representative group-numbering. In other words, when we search for the neighbors of a partitioning $\mathcal{A}_\square$, we can use any representative group-numbering (for example $\mathcal{A}$); then we find the neighbor $\mathcal{A}^\square$ of $\mathcal{A}$ by COG; $\mathcal{A}^\square$ is a representative of $\mathcal{A}_\square^\square$, which is the wanted neighbor of $\mathcal{A}_\square$.

Fig. 3. The commutative diagram.

x to move, the new part p. Let us notice that the third parameter is linked to the first one: part p must be a part of the partitioning $\mathcal{A}_\sim$. Technically, we can operate on representative group-numberings. Therefore, we use the classical operator, *change_object_group* (COG), which consists in changing the group of an object. Intuitively, the chosen representative group-numbering simulates the whole equivalence class. Figure 3 summarizes this idea which is proved hereafter.

Proof. Diagram commutativity.

Let $\mathcal{A}_\sim$ be a partitioning represented by $\mathcal{A}_1$ and $\mathcal{A}_2$. There exists a bijection σ such that $\forall x \in U, \sigma(\mathcal{A}_1(x)) = \mathcal{A}_2(x)$. Let x be an object and let i be a group of $\mathcal{A}_1$. Let $\mathcal{A}_1'$ (resp. $\mathcal{A}_2'$) be the group-numbering obtained by $\mathcal{A}_1' = COG(\mathcal{A}_1, x, i)$ (resp. $\mathcal{A}_2' = COG(\mathcal{A}_2, x, \sigma(i))$). For all $y \neq x$, $\sigma(\mathcal{A}_1'(y)) = \sigma(\mathcal{A}_1(y)) = \mathcal{A}_2(y) = \mathcal{A}_2'(y)$ and $\sigma(\mathcal{A}_1'(x)) = \sigma(i) = \mathcal{A}_2'(x)$. So, σ is effectively a bijection with regard to the definition of $\sim$. Thus $\mathcal{A}_1' \sim \mathcal{A}_2'$, and consequently the diagram is commutative.

4 Metrics of Partitioning Set

In this section, we define a new mathematical object, namely a refining matrix. Then we use this object to prove that there is a metric on partitioning space, that can be computed polynomially. Thanks to such a metric, one can use some diversification mechanism or analyzes the fitness landscape of his problem.

4.1 Refining Matrix

When we want to estimate the distance in the partitioning space, we are tempted to use the distance d_n between group-numberings defined by:
$d_n(\mathcal{A}, \mathcal{B}) = \sum_{x \in U} 1 - \delta_{\mathcal{A}(x)\mathcal{B}(x)}$, where δ_{ij} is the Kronecker symbol ($= 1$, if i $=$ j; 0, otherwise). However due to the symmetry, this distance may provide great value between two group-numberings representing the same partitioning.

To solve this problem, we propose a metric on the set of partitioning, that we call d_σ. The metric is based on the use of a dedicated object: the refining matrix (of two group-numberings). Such a matrix $\mathcal{M}^{AB} = (m_{ij}^{AB})$, for any group-numberings $\mathcal{A}$ and $\mathcal{B}$, is defined as $m_{ij}^{AB} = \#(\mathcal{A}^{-1}(i) \cap \mathcal{B}^{-1}(j))$; where $\#$ means the cardinality of a set.

We also define the indicator e associated with the refining matrix as:

$$\mathcal{R} \times \mathcal{R} \times \{\{1, 2, \ldots N\} \to \{1, 2, \ldots N\}\} \to \mathbb{N}$$
$$\mathcal{A}, \quad \mathcal{B}, \qquad\qquad\qquad \tau \qquad\qquad \mapsto e(\mathcal{A}, \mathcal{B}, \tau) = N - \sum_i m_{i, \tau(i)}^{AB}$$

We have trivially the following property:

$$(e(\mathcal{A}, \mathcal{B}, \tau) = 0) \Leftrightarrow (\forall i, j \in \{1, 2, \ldots N\}), (j \neq \tau(i) \Rightarrow m_{i,j}^{AB} = 0) \qquad (1)$$

Indeed, $e(\mathcal{A}, \mathcal{B}, \tau) = 0$) can be seen as the sum of non-negative terms $(\sum_i \sum_{j \neq \tau(i)} m_{i,j}^{AB})$. And this sum is null if and only if each of the terms is null.

4.2 Metrics d_σ

The mapping τ can be seen as a relabeling of the group of a group-numbering. So, it is about finding one of the best bijections σ. Formally, for two given group-numberings $\mathcal{A}$ and $\mathcal{B}$, we search a bijection σ^{AB}, so that $e(\mathcal{A}, \mathcal{B}, \sigma^{AB})$ is minimal. The distance d_σ on partitionings is defined as $d_\sigma(\mathcal{A}_\sim, \mathcal{B}_\sim) = e(\mathcal{A}, \mathcal{B}, \sigma^{AB})$.

We have to verify four properties to prove that d_σ is a distance:

1. The definition of d_σ is independent of the choice of representatives and the (eventual) choice of minimal σ^{AB}. In other words, d_σ is effectively defined.
2. d_σ must be symmetrical,
   i.e. for any $\mathcal{A}_\sim$ and $\mathcal{B}_\sim$, $d_\sigma(\mathcal{A}_\sim, \mathcal{B}_\sim) = d(\mathcal{B}_\sim, \mathcal{A}_\sim)$.
3. d_σ must be separated,
   i.e. $d_\sigma(\mathcal{A}_\sim, \mathcal{B}_\sim) = 0$ is equivalent to $\mathcal{A}_\sim = \mathcal{B}_\sim$.
4. d_σ must verify the triangular inequality,
   i.e $d_\sigma(\mathcal{A}_\sim, \mathcal{C}_\sim) \leq d_\sigma(\mathcal{A}_\sim, \mathcal{B}_\sim) + d_\sigma(\mathcal{B}_\sim, \mathcal{C}_\sim)$.

Proof. d_σ is effectively defined.

Let $\mathcal{A}$ and $\mathcal{B}$ be two group-numberings. Let σ and σ' be two bijections verifying the definition of d_σ. So, $e(\mathcal{A}, \mathcal{B}, \sigma)$ is minimal among bijections. Since σ' is a bijection, $e(\mathcal{A}, \mathcal{B}, \sigma) \leq e(\mathcal{A}, \mathcal{B}, \sigma')$. For the same reason, we have $e(\mathcal{A}, \mathcal{B}, \sigma') \leq e(\mathcal{A}, \mathcal{B}, \sigma)$. Thus $e(\mathcal{A}, \mathcal{B}, \sigma) = e(\mathcal{A}, \mathcal{B}, \sigma')$. Hence, the computation of $e(\mathcal{A}, \mathcal{B}, \sigma^{AB})$ is independent of the choice of σ^{AB}.

Let $\mathcal{A}$ (resp. $\mathcal{B}$) and $\mathcal{A}'$ (resp. $\mathcal{B}'$) be two representatives of $\mathcal{A}_\sim$ (resp. $\mathcal{B}_\sim$). $\mathcal{A}$ is equivalent to $\mathcal{A}'$, thus there exists σ_1 (resp. σ_2) so that $\forall x \in U$, $\mathcal{A}'(x) = \sigma_1(\mathcal{A}(x))$ (resp. $\mathcal{B}'(x) = \sigma_2(\mathcal{B}(x))$). We have obviously $m_{ij}^{A'B'} = m_{\sigma_1(i)\sigma_2(j)}^{AB}$. So, $e(\mathcal{A}', \mathcal{B}', \sigma^{A'B'}) = N - \sum_i m_{\sigma_1(i)\sigma_2(\sigma^{A'B'}(\sigma_1(i)))}^{AB}$, which is naturally greater or equal than $e(\mathcal{A}, \mathcal{B}, \sigma^{AB})$. Since $\mathcal{A}$ (resp. $\mathcal{B}$) and $\mathcal{A}'$ (resp. $\mathcal{B}'$) play the same role, we also have: $e(\mathcal{A}, \mathcal{B}, \sigma^{\mathcal{A}\mathcal{B}}) \geq e(\mathcal{A}'\mathcal{B}', \sigma^{\mathcal{A}'\mathcal{B}'})$. Hence $e(\mathcal{A}\mathcal{B}, \sigma^{\mathcal{A}\mathcal{B}}) = e(\mathcal{A}', \mathcal{B}', \sigma^{\mathcal{A}'\mathcal{B}'})$. Consequently, d_σ is effectively defined independently from the choice of representatives.

As a conclusion, the definition of d_σ is not ambiguous.

Proof. d_σ is symmetrical.

Let $\mathcal{A}_\sim$ and $\mathcal{B}_\sim$ be two partitionings. Let $\mathcal{A}$ and $\mathcal{B}$ be the respective associated representatives. We know that $e(\mathcal{A}, \mathcal{B}, \sigma^{\mathcal{A}\mathcal{B}}) = N - \sum_i m^{\mathcal{A}\mathcal{B}}_{i, \sigma^{\mathcal{A}\mathcal{B}}(i)} = D$ is minimal. $e(\mathcal{B}, \mathcal{A}, \sigma^{\mathcal{A}\mathcal{B}^{-1}}) = N - \sum_i m^{\mathcal{B}\mathcal{A}}_{i, \sigma^{\mathcal{A}\mathcal{B}^{-1}}(i)}$. Since, $M^{\mathcal{B}\mathcal{A}}$ is the transposed matrix of $M^{\mathcal{A}\mathcal{B}}$, we can deduce $e(\mathcal{B}, \mathcal{A}, \sigma^{\mathcal{A}\mathcal{B}^{-1}}) = N - \sum_i m^{\mathcal{A}\mathcal{B}}_{\sigma^{\mathcal{A}\mathcal{B}^{-1}}(i), i}$. By changing the index $j = \sigma^{-1}(i)$, this leads to $e(\mathcal{B}, \mathcal{A}, \sigma^{\mathcal{A}\mathcal{B}^{-1}}) = N - \sum_{\sigma^{\mathcal{A}\mathcal{B}}(j)} m^{\mathcal{A}\mathcal{B}}_{j, \sigma^{\mathcal{A}\mathcal{B}}(j)}$. Since $\sigma^{\mathcal{A}\mathcal{B}}(j)$ traverses all indexes, we can simplify the term under the sum. We obtain:

$$e(\mathcal{B}, \mathcal{A}, \sigma^{\mathcal{A}\mathcal{B}^{-1}}) = N - \sum_j m^{\mathcal{A}\mathcal{B}}_{j, \sigma^{\mathcal{A}\mathcal{B}}(j)} = D \qquad (2)$$

Let σ' be a bijection. By using the same method as before, we deduce that $e(\mathcal{B}, \mathcal{A}, \sigma') = e(\mathcal{A}, \mathcal{B}, \sigma'^{-1}) \geq e(\mathcal{A}, \mathcal{B}, \sigma^{\mathcal{A}\mathcal{B}}) = e(\mathcal{B}, \mathcal{A}, \sigma^{\mathcal{A}\mathcal{B}^{-1}})$. Thus, $e(\mathcal{B}, \mathcal{A}, \sigma^{\mathcal{A}\mathcal{B}^{-1}})$ is effectively minimal and equals to D). Consequently, $d_\sigma(\mathcal{A}, \mathcal{B}) = d_\sigma(\mathcal{B}, \mathcal{A})$.

Proof. $d_\sigma(\mathcal{A}_\sim, \mathcal{B}_\sim) = 0 \Longrightarrow \mathcal{A}_\sim = \mathcal{B}_\sim$.

Let $\mathcal{A}_\sim$ and $\mathcal{B}_\sim$ be two partitionings. Let $\mathcal{A}$ (resp. $\mathcal{B}$) be a representative of $\mathcal{A}_\sim$ (resp. $\mathcal{B}_\sim$). Let us assume that $d_\sigma(\mathcal{A}_\sim, \mathcal{B}_\sim) = 0$, i.e. $e(\mathcal{A}, \mathcal{B}, \sigma^{\mathcal{A}\mathcal{B}}) = 0$.

Let us reason by contradiction, by assuming there exists an object x verifying $\sigma^{\mathcal{A}\mathcal{B}}(\mathcal{A}(x)) \neq \mathcal{B}(x)$. Consequently, $m^{\mathcal{A}\mathcal{B}}_{\mathcal{A}(x)\mathcal{B}(x)} \geq 1$. This contradicts the assumption of null distance. So, such an object cannot exist. Hence, we have for all objects x, $\sigma^{\mathcal{A}\mathcal{B}}(\mathcal{A}(x)) = \mathcal{B}(x)$. Thus $\mathcal{A}$ is equivalent to $\mathcal{B}$ (i.e. $\mathcal{A}_\sim = \mathcal{B}_\sim$).

Proof. $d_\sigma(\mathcal{A}_\sim, \mathcal{B}_\sim) = 0 \Longleftarrow \mathcal{A}_\sim = \mathcal{B}_\sim$.

Let $\mathcal{A}_\sim$ be a partitioning. $\mathcal{A}_\sim$ can be represented by two group-numberings $\mathcal{A}$ and $\mathcal{B}$. By definition, $\mathcal{A} \sim \mathcal{B}$. There exists a bijection $\sigma : \{1, 2, \ldots N\} \to \{1, 2, \ldots N\}$, so that for all objects $x \in U$, $\sigma(\mathcal{A}(x)) = \mathcal{B}(x)$ is verified. Consequently, the refining matrix of $\mathcal{A}$ and $\mathcal{B}$ contains at most one non-zero number in each line (and each column); These elements are $m_{i,\sigma(i)}$. By using equation (1), we obtain $e(\mathcal{A}, \mathcal{B}, \sigma^{\mathcal{A}\mathcal{B}}) = 0$. Thus $d_\sigma(\mathcal{A}_\sim, \mathcal{B}_\sim) = 0$.

To prove the triangular inequality, we use the two following lemmas.

Lemma 1. *For any group-numberings $\mathcal{A}$ and $\mathcal{B}$, and any neighbor $\mathcal{A}'$ of $\mathcal{A}$, we have the following relation: $|e(\mathcal{A}, \mathcal{B}, \sigma^{\mathcal{A}\mathcal{B}}) - e(\mathcal{A}', \mathcal{B}, \sigma^{\mathcal{A}'\mathcal{B}})| \leq 1$.*

Lemma 2. *For any group-numberings $\mathcal{A}$ and $\mathcal{B}$, for all objects x verifying $\sigma^{\mathcal{A}\mathcal{B}}(\mathcal{A}(x)) \neq \mathcal{B}(x)$. We have:*
$$e(\mathcal{A}, COG(\mathcal{B}, x, \sigma^{\mathcal{A}\mathcal{B}}(\mathcal{A}(x))), \sigma^{\mathcal{A}\ COG(\mathcal{B}, x, \sigma^{\mathcal{A}\mathcal{B}}(\mathcal{A}(x)))}) - e(\mathcal{A}, \mathcal{B}, \sigma^{\mathcal{A}\mathcal{B}}) = -1.$$

Proof. of triangular inequality

Let $\mathcal{A}$ and $\mathcal{C}$ be two group-numberings. Let us show by recurrence for all $\mathcal{B}$, $e(\mathcal{A}, \mathcal{C}, \sigma^{\mathcal{AC}}) \le e(\mathcal{A}, \mathcal{B}, \sigma^{\mathcal{AB}}) + e(\mathcal{B}, \mathcal{C}, \sigma^{\mathcal{BC}})$.

1. Let $\mathcal{P}_n$ be the following statement: "$\forall \mathcal{B}$ so that $e(\mathcal{A}, \mathcal{B}, \sigma^{\mathcal{AB}}) = n$", then we have the following inequality: $e(\mathcal{A}, \mathcal{C}, \sigma^{\mathcal{AC}}) \le e(\mathcal{A}, \mathcal{B}, \sigma^{\mathcal{AB}}) + e(\mathcal{B}, \mathcal{C}, \sigma^{\mathcal{BC}})$.
2. Since d_σ is separated, $\mathcal{P}_0$ is obviously true.
3. Let us assume $\mathcal{P}_n$. Let $\mathcal{B}$ be a group-numbering such that $d_\sigma(\mathcal{A}_\sim, \mathcal{B}_\sim) = n+1$. There exists an object x such that $A(x) \neq \sigma^{\mathcal{AB}}(A(x))$. Let us build $\mathcal{B}' = COG(\mathcal{B}, x, \sigma^{\mathcal{AB}}(A(x)))$. By lemma 2, we have $e(\mathcal{A}, \mathcal{B}', \sigma^{\mathcal{AB}'}) = n$. So, we can deduce by the recurrence assumption that $e(\mathcal{A}, \mathcal{C}, \sigma^{\mathcal{AC}}) \le e(\mathcal{A}, \mathcal{B}', \sigma^{\mathcal{AB}'}) + e(\mathcal{B}', \mathcal{C}, \sigma^{\mathcal{B}'\mathcal{C}})$. Now, lemma 1 gives that $e(\mathcal{B}, \mathcal{C}, \sigma^{\mathcal{BC}}) \ge e(\mathcal{B}', \mathcal{C}, \sigma^{\mathcal{B}'\mathcal{C}}) - 1$.

$$e(\mathcal{A}, \mathcal{B}, \sigma^{\mathcal{AB}}) + e(\mathcal{B}, \mathcal{C}, \sigma^{\mathcal{BC}}) \ge n + 1 + e(\mathcal{B}', \mathcal{C}, \sigma^{\mathcal{B}'\mathcal{C}}) - 1$$
$$e(\mathcal{A}, \mathcal{B}, \sigma^{\mathcal{AB}}) + e(\mathcal{B}, \mathcal{C}, \sigma^{\mathcal{BC}}) \ge n + e(\mathcal{B}', \mathcal{C}, \sigma^{\mathcal{B}'\mathcal{C}})$$
$$e(\mathcal{A}, \mathcal{B}, \sigma^{\mathcal{AB}}) + e(\mathcal{B}, \mathcal{C}, \sigma^{\mathcal{BC}}) \ge e(\mathcal{A}, \mathcal{B}', \sigma^{\mathcal{AB}'}) + e_\sigma(\mathcal{B}', \mathcal{C}, \sigma^{\mathcal{B}'\mathcal{C}})$$
$$e(\mathcal{A}, \mathcal{B}, \sigma^{\mathcal{AB}}) + e(\mathcal{B}, \mathcal{C}, \sigma^{\mathcal{BC}}) \ge e(\mathcal{A}, \mathcal{C}, \sigma^{\mathcal{AC}})$$
$$d(\mathcal{A}_\sim, \mathcal{B}_\sim) + d_\sigma(\mathcal{B}_\sim, \mathcal{C}_\sim) \ge d_\sigma(\mathcal{A}_\sim, \mathcal{C}_\sim)$$

4. So d_σ verify the triangular inequality.

Proof. of lemma 1

Let $\mathcal{A}$ and $\mathcal{B}$ be two group-numberings. Let $\mathcal{A}'$ be a neighbor of $\mathcal{A}$ by changing the group of object x. Let $\mathcal{M}^{\mathcal{AB}}$ (resp. $\mathcal{M}^{\mathcal{A}'\mathcal{B}}$) be the refining matrix induced by the computation of $e(\mathcal{A}, \mathcal{B}, \sigma^{\mathcal{AB}})$ (resp. $e(\mathcal{A}', \mathcal{B}, \sigma^{\mathcal{A}'\mathcal{B}})$). We have the four following relations:

$$|\sum_i m_{i\sigma^{\mathcal{AB}}(i)}^{\mathcal{AB}} - \sum_i m_{i\sigma^{\mathcal{AB}}(i)}^{\mathcal{A}'\mathcal{B}}| \le 1 \tag{3}$$

Because changing the refining matrix is equivalent to decrement one element and increment another one of the same line.

$$|\sum_i m_{i\sigma^{\mathcal{A}'\mathcal{B}}(i)}^{\mathcal{A}'\mathcal{B}} - \sum_i m_{i\sigma^{\mathcal{A}'\mathcal{B}}(i)}^{\mathcal{AB}}| \le 1 \tag{4}$$

For the same reason.

$$\sum_i m_{i\sigma^{\mathcal{AB}}(i)}^{\mathcal{AB}} - \sum_i m_{i\sigma^{\mathcal{A}'\mathcal{B}}(i)}^{\mathcal{AB}} \ge 0 \tag{5}$$

Because $\sigma^{\mathcal{AB}}$ is minimal for $\mathcal{M}^{\mathcal{AB}}$

$$\sum_i m_{i\sigma^{\mathcal{A}'\mathcal{B}}(i)}^{\mathcal{A}'\mathcal{B}} - \sum_i m_{i\sigma^{\mathcal{AB}}(i)}^{\mathcal{A}'\mathcal{B}} \ge 0 \tag{6}$$

Because $\sigma^{\mathcal{A}'\mathcal{B}}$ is minimal for $\mathcal{M}^{\mathcal{A}'\mathcal{B}}$.

From this relation, we deduce:

$$\sum_i m^{AB}_{i\sigma^{AB}(i)} - \sum_i m^{A'B}_{i\sigma^{AB}(i)} \leq 1$$
$$\sum_i m^{AB}_{i\sigma^{AB}(i)} - \sum_i m^{A'B}_{i\sigma^{AB}(i)} \leq 1 + \sum_i m^{A'B}_{i\sigma^{A'B}(i)} - \sum_i m^{A'B}_{i\sigma^{AB}(i)}$$
$$\sum_i m^{AB}_{i\sigma^{AB}(i)} - \sum_i m^{A'B}_{i\sigma^{A'B}(i)} \leq 1$$

and

$$\sum_i m^{A'B}_{i\sigma^{A'B}(i)} - \sum_i m^{AB}_{i\sigma^{A'B}(i)} \leq 1$$
$$\sum_i m^{A'B}_{i\sigma^{A'B}(i)} - \sum_i m^{AB}_{i\sigma^{A'B}(i)} \leq 1 + \sum_i m^{AB}_{i\sigma^{AB}(i)} - \sum_i m^{AB}_{i\sigma^{A'B}(i)}$$
$$\sum_i m^{A'B}_{i\sigma^{A'B}(i)} - \sum_i m^{AB}_{i\sigma^{AB}(i)} \leq 1$$

So, $|\sum_i m^{AB}_{i\sigma^{AB}(i)} - \sum_i m^{A'B}_{i\sigma^{A'B}(i)}| \leq 1$;
i.e. $|e(A, B, \sigma^{AB}) - e(A', B, \sigma^{A'B})| \leq 1$.

Proof. of lemma 2

Let A and B be two group-numberings. Let x be an object verifying $\sigma^{AB}(A(x)) \neq B(x)$. Let us state $B' = COG(B, x, \sigma^{AB}(A(x)))$. We notice that σ^{AB} can be chosen as maximal bijection for $\sigma^{AB'}$. Indeed, the element of the refining matrix of index $A(x)\sigma^{AB}(A(x))$ increased while $A(x)B(x)$ decreased. The other elements of the matrix stay unchanged : $m_{ij} = \#(A^{-1}(i) \bigcap B^{-1}(j)) = \#(A^{-1}(i) \bigcap B'^{-1}(j))$. So, we have $e(A, B', \sigma^{AB}) - e(A, B, \sigma^{AB}) = -1$

We proved theoretically that d_σ is a metric. Now, we need to compute its value. We have to find a bijection σ verifying the definition. Fortunately, finding a minimal permutation σ is equivalent to solve a linear assignment problem which can be polynomially done with the Hungarian algorithm [1].

5   Conclusion and Future Works

In this paper, we show how to break the symmetry of the partitioning space. We propose a theoretical construction of the partitioning space that allows us to define a test of equality between partitionings, which is computable in $O(N)$. We keep the classic/elementary neighborhood operator as simple (in computational time) than it was.

After that, build a mathematical tool namely the refining matrix and we use it to show that it is possible to estimate the distance between partitionings, in mathematical sense. Finally, we state that the proposed metric is polynomially computable. By the way, we obtain tools allowing to use some classic niching mechanisms or landscape studies.

We are now investigating new applications of the refining matrix to design new tools to operate on partitioning sets such as genetic operators.

References

1. G. Carpaneto and P. Toth. Algorithm 548: Solution of the assignment problem. *ACM Transactions on Mathematical Software (TOMS)*, 6(1):104–111, 1980.
2. E. Falkenauer. *Genetic Algorithm and Grouping Problems*. John Wiley & Sons, 1998.
3. P. Galinier and J-K. Hao. Hybrid evolutionary algorithms for graph coloring. *Journal of Combinatorial Optimization*, 3(4):379–397, 1999.
4. S. Hurley, D. Smith, and C. Valenzuela. A permutation based genetic algorithm for minimum span frequency assignment. In *PPSN V*, volume 1498 of *LNCS*, pages 907–916, Amsterdam, Sept 1998. Springer-Verlag Publication.
5. S. Mahfoud. *Niching Methods for Genetic Algorithm*. PhD thesis, Universty of Illinois, 1995.
6. A. Marino and R. I. Damper. Breaking the symmetry of the graph colouring problem with genetic algorithms. In Darrell Whitley, editor, *Late Breaking Papers at the 2000 Genetic and Evolutionary Computation Conference*, pages 240–245, Las Vegas, Nevada, USA, april 2000.
7. E.D. Weinberger. Correlated and uncorrelated fitness landscapes and how to tell the difference. *Biological Cybernetics*, pages 325–336, 1990.

Author Index